TIME BOMB

Books by Naomi A. Hintze

YOU'LL LIKE MY MOTHER

THE STONE CARNATION

ALOHA MEANS GOODBYE

LISTEN, PLEASE LISTEN

CRY WITCH

BURIED TREASURE WAITS FOR YOU

THE PSYCHIC REALM: WHAT CAN YOU BELIEVE? *(Co-author)*

Peter van der Linde
with Naomi A. Hintze

TIME BOMB

LNG: The truth about our newest and most
dangerous energy source

Doubleday & Company, Inc., Garden City, New York, 1978

Material that appeared originally in The New York *Times* is used by permission of The New York Times Company.

Excerpt from "Has It Been 20 Years?" by George E. Condon, which appeared originally in *The Plain Dealer*, is reprinted with permission.

Grateful acknowledgment is made for the use of material from *Supership* by Noël Mostert. Copyright © 1976 by Noël Mostert. Published by Alfred A. Knopf, Inc.

Library of Congress Cataloging in Publication Data

Van der Linde, Peter, 1949–
Time bomb.

Bibliography
1. Liquefied natural gas—Transportation.
I. Hintze, Naomi A., joint author. II. Title.
TP761.L5V36   665'.74
ISBN: 0-385-12979-3
Library of Congress Catalog Card Number 77-76271

To Linda

TIME BOMB

"THE FINAL MINUTES"

(Prologue)

5:05 P.M. First Mate Brian Miller stands on the bridge of the *Bonnie Smathers,* a 50,000-ton tanker loaded with oil from the Persian Gulf. She enters the Arthur Kill, the waterway between Staten Island and the New Jersey shore.

Brian is not apprehensive—he has navigated this channel many times—but he is alert because the Arthur Kill has six of the eight Coast Guard-designated hazards of New York Harbor. Two ninety-degree turns lie ahead, first right, then left, in quick succession.

The pilot, who has boarded at Ambrose Light, sits to the first mate's left in a swivel chair, feet up, as he sips a cup of coffee and looks through the wide plate-glass windows that span the front of the navigating bridge. This run is routine for him. He knows every bend, each submerged and semi-submerged obstruction that lies outside the red and black marker buoys outlining the 500-foot-wide navigable channel. Immediately behind the two men is the helmsman at the steering stand.

On the forward deck, four stories below, final preparations have begun for docking at the terminal six miles ahead. The windlass is bringing up the mooring hawsers, thick as the big bos'n's arm, and the deck gang is "faking" them out on the fo'c'sle head.

It is warm for early May. The grounds of the Mount

Loretto orphanage to the right show patches of green grass, a few trees struggling into bloom. But any scents that might have brought a welcome whiff of spring are canceled by the murk of fumes that spew from the proliferating smokestacks on both sides of the Kill. Here within a very few miles is the concentration of plants, refineries, and tank farms that supply a large percentage of the petrochemical needs of the entire United States.

Brian's hands are easy on the throttles, his eyes flicking over some two hundred knobs and lights and gauges and levers and buttons on the control console before him. A glance at the rudder-angle indicator in the center of the board shows that the steering is responding smoothly to each course change; both engines are in sync, running at 90 r.p.m.; the vessel's speed is fourteen knots; and the northeasterly wind is a variable five to six miles per hour.

The *Bonnie Smathers* is a sweet ship; Brian knows her well and has been on her for six months. But he has "channel fever" now and is headed for home, thank God. Six months is a long time to be away from a girl like Cathy. They have been married a year and when he left they hadn't been altogether sure she was pregnant.

To his right between the intervening structures, Brian catches intermittent glimpses of the pale green fourteen-story liquefied natural gas storage tanks. Pluperfect Gas Corporation owns them. More than four years of local protests prevented their use, but now each tank holds 37,800,000 gallons of liquefied natural gas. Incredible stuff, that LNG. Chilled to 260 degrees below zero Fahrenheit, it becomes six hundred times more concentrated than the gas that comes from the wells.

The ships that carry LNG are almost a thousand feet long. Somewhere Brian has read that, in order to carry that much energy in its gaseous state, the carriers would have to be more than a hundred *miles* long.

Wow. Staggers the imagination just to think of it. The

ship would be longer than Long Island! Glad not to be sailing on an LNG carrier. Glad he and Cathy had bought their house a good safe three miles away from the tanks because if even one of them ever let go . . . well, what the hell, why let yourself think about it? The tanks were considered—how was it they put it?—an "acceptable risk." Something like that. Not only Pluperfect Gas but the Federal Power Commission, the Coast Guard, everybody, except for a few worry warts, thought they were safe.

Not to worry. That's what he told Cathy.

5:07. A Little League game is in progress. Mothers cheer the players, boo the umpire. Small sisters and brothers, giving scant attention to the game, chase each other, fight, fall down, bawl, run and play some more. A few hundred feet to the west, the towering LNG tanks block the afternoon sun, casting a welcome shadow across the athletic field.

The big flap occasioned by the construction and filling of the tanks is pretty well forgotten now as so much hysteria. After all, they have been in operation for nearly two years and nothing has happened. And . . . look at it this way, the government wouldn't have let them be put there if there had been any real danger—right? Right. . . .

5:10. A garbage scow passes the *Bonnie Smathers* heading out to sea to dump beyond Ambrose Light. This is the first vessel of any size that Brian has seen since entering the channel. He comments on the light traffic and the pilot tells him, "There's an LNG carrier, *Global Glory*, about to offload at Smoking Point. Coast Guard, y'know, has to clear the channel of all craft of any size when she's escorted in. We'll see plenty of traffic coming along now, trying to make up for lost time. Gripes hell out of 'em to have to wait for those big mothers. But Coast Guard regulations—what can they do?"

5:15. The *Bonnie Smathers* is approaching the first ninety-degree bend to the right. The pilot has given the order to slow down to half speed ahead. Brian is a little concerned about the falling tide. His ship is sniffing the bottom and squatting as it is drawn down closer to the mud. Not that this is anything unusual; in shallow water and at reduced speed a ship is always less maneuverable.

The captain bats back the plastic curtain between the chart room and the navigating bridge and approaches the control console. He is a humorless, unapproachable man, but professional, a seaman of the first order, honed through decades of experience. His critical gaze takes in the console, scanning for the danger signals of a flashing light or a flickering needle that might indicate malfunction. He picks up his binoculars and puts them to his eyes without checking the adjustment; nobody else ever touches the "old man's" glasses.

The pilot pulls his feet down from the window sill and leans forward in his chair as the *Bonnie Smathers* approaches her first turn to the right.

"Fifteen degrees right rudder."

"Fifteen right," repeats the helmsman.

"Hard right rudder." It is necessary to increase her rate of turn because of the sharpness of the bend.

"Hard right," the helmsman echoes the command.

The needle of the rudder-angle indicator sweeps to the right, duplicating the movement of the vessel's rudder. Tick, tick, tick, the gyrocompass can be heard as the bow pans the shore. And the shore installations file past as if they were moving and not the *Bonnie Smathers*.

Well into the right turn, the pilot says, "Midships the rudder."

"Rudder is midships," the helmsman replies.

It flashes through Brian's mind that the pilot may have misjudged the turn by a couple of seconds, shouldn't have stayed so long with hard right since another turn, ninety degrees to the left, lies just ahead.

But it is unwritten law that nobody except the captain ever voices an opinion contrary to the pilot's commands. Brian shoots a questioning look at the captain, noting the knotting of his jaw muscles. The continued rapid clicking of the gyrocompass announces that the turn to the right has not yet been broken. The vessel's bow is still swinging fast to the right.

Pilot: "Hard left rudder."

Helmsman: "Hard left." He spins the steering wheel to the left.

Now, although they are still separated by half a mile, the monstrous LNG carrier is visible, tied up on the outside corner of the next bend to the left. Brian is sure the *Bonnie Smathers* can make it out of the turn in time, but his gut tightens and his hands begin to sweat.

On the *Global Glory* the deck hands are connecting the cargo loading arms, wrestling the flanges into alignment in the always difficult task of slipping that first bolt that will bond the ship and shore cargo lines together. They pay no attention to the approaching tanker although they can feel the *Global Glory* scrape against the dock in response to the displaced water that precedes the arrival of a large ship.

5:20. The bow of the *Bonnie Smathers* is still swinging fast to the right in spite of the pilot's command for hard left. He shouts now, "Hard left!"

Helmsman: "I've got her hard left."

Captain: "She's not breaking that turn to the right."

The pilot, between his teeth: "I know, I *know*."

The helmsman's widened eyes dart from pilot to captain to Brian.

The *Bonnie Smathers* is not responding. Bank suction, that invisible force beneath the water, causes her bow to move to starboard as if drawn by a magnet.

There is no way under heaven to stop a laden ship of this

size in so short a space. She has too much momentum. There is only one responsible option.

"Emergency full ahead!"

Brian slams the throttles forward even before he repeats the ahead order. This is good seamanship, trying to power out of it by pouring it on, a last desperate attempt to break the continued right swing.

The turbochargers scream for air under the strain of this sudden demand for increased speed. A chaos of alarms caused by the overloading of all engine systems is unnoticed by the four men on the bridge. Their gaze is riveted on the *Global Glory* two ship lengths ahead.

Down in the engine room the crew has no way of knowing what is taking place, but on the forward deck the men have dropped the lines and are running aft.

The gyrocompass groans as the turn to the right finally breaks. It may be soon enough. There is still a chance. Brian allows himself a gulp of relief. He hears himself coaxing the ship under his breath, "Start left, come *on, start left!*" He has been in perilous situations many times before and has not allowed himself to think that it is all over. But he knows that this time, even if they miss collision with that terrible LNG ahead, their suction may pull her away from the dock, snapping her mooring cables.

The captain grabs the whistle cord, pulls it down, giving the rapid short blasts of warning required by the *Rules of the Road*. The *Bonnie Smathers* is bucking wildly, her stern jerking up and down as the propellers dig into the mud of the river bottom. It is almost impossible for the men on the bridge to keep their footing. They clutch for support, brace themselves, pray that the rapidly lessening space will let them ease by, scraping perhaps, but . . .

5:22. The captain's cry of "Mayday! Mayday!" is heard on all radiotelephone stations seconds before the bow of the *Bonnie Smathers* rips at full speed ahead into the mid-section of the *Global Glory*'s hull, splinters the steel of one of the five LNG cargo tanks.

Ten thousand tons of liquefied natural gas gush out, colder than the dark side of the moon. It covers everything, everybody. It fractures the deck of the *Bonnie Smathers* and instantly freezes the surrounding waters of the Kill. Trillions of ice crystals boil and burst, forming an expanding white vapor cloud.

Silently the vapor flows, hugging the surface, asphyxiating, quick-freezing all in its path. Warmed by ambient temperatures, wind-borne, the vapor becomes a low-lying plume, pale and pretty, and only the few knowledgeable viewers who stare from a distance suspect its sinister potential.

The plume searches for an ignition source. Any spark will suffice, a cigarette, a pilot light, a back-yard barbecue. No one will ever know what sets off the fire, the instantaneous tongue of flame that races back to its source, incinerating everything in its course.

It reaches the breached *Global Glory*. Explosions occur in series as the ship's other four cargo tanks rupture and their vapors ignite. Harbor traffic nearby melts in the blast-furnace heat, twists, catapults skyward.

Shards of flying steel penetrate the Pluperfect LNG shore storage tanks, and tens of millions of gallons of LNG pour out and down like a tidal wave, vaporizing. Seconds later ignition occurs. Shock waves flatten oil and gasoline storage tank farms and in moments the Kill is an engulfing wall of flame. Thousands of chemical tanks contribute their noxious contents to the brew.

Smoke and flying debris eclipse the sun and winds of hurricane force suck, spread, and within the ten-mile reach of the spiraling arms, no one lives. No one . . .

A motorist on the Long Island Expressway slams on his brakes and slews crazily to the outer lane to stare in disbelief toward the southwestern horizon. The skyline alters, lifts in hideous rearrangement as tall structures blow apart. The jolting shocks feel like an earthquake. It looks like the end of the world.

That enormous fireball must mean a nuclear explosion. *Oh, God, we are at war.*

5:35. A news flash halts a comedy rerun on NBC: "We interrupt our broadcast to bring you this bulletin: An explosion of undetermined origin has rocked Staten Island. Many are feared dead and property damage is believed to be extensive. Stay tuned to this station for more news as it develops."

The charnel zone extends beyond the uncountable acres of total demolition. People who can still move do so aimlessly, some of them running, not knowing where to go to escape the encroaching fires and toppling buildings. Shattered glass, blocks of stone, girders, dismembered bodies rain from the sky. Planes plummet to the ground like dead birds.

Eardrums are split by the continuing explosions. No one hears the groans of the dying crushed under the rubble. Tripped burglar alarms, bells, sirens are a cacophony fit for Dante's Inferno. Flames are reflected on the face of a woman who stumbles, staring, a dead child in her arms.

6:31. Evening News: "Staten Island has been virtually destroyed. The New Jersey coastal area from Perth Amboy to Bayonne lies in ruins. But this is not an atomic attack as was first reported. Repeat, this is *not* an atomic attack.

"The initial cause, according to information available at this time, seems to have been a collision at about 5:30, Eastern Daylight Time, between an oil tanker and the *Global Glory*, a liquefied natural gas carrier, at Smoking Point on the Arthur Kill. The subsequent rupturing of the Pluperfect Gas Corporation's nearby tanks storing millions of gallons of

LNG seems to have been primarily responsible for the enormity of the disaster.

"In a domino effect, hundreds of refineries, petrochemical tanks, and other storage facilities on the waterway have exploded and are burning in an unprecedented holocaust.

"A bulletin has just been handed to me: Newark International Airport is in flames, together with an estimated hundred or more commercial jet liners.

"Loss of life, for which estimates range upwards of a million, and property damage into the billions place this disaster as unquestionably the worst this nation has ever known and one of the worst single catastrophes in the recorded history of the world.

"A reporter who viewed the scene from the top of the World Trade Center half an hour ago said, 'I saw Hiroshima. Compared to this, it was nothing.'

"Fires are spreading to adjacent areas and all available fire-fighting and rescue units are rushing to render aid to the thousands of injured. Highways are jammed as residents flee homes that are still endangered. All bridges to Staten Island are out.

"Here in midtown Manhattan, approximately fifteen miles from the Staten Island LNG facility, the effects are being felt and there is a great deal of panic. We urge those of you who have not been injured to stay in your homes to allow emergency vehicles access to affected areas.

"We have been told that the governors of New York and New Jersey are in conference. It is expected that they will issue statements shortly.

"The executives of Pluperfect Gas Corporation, owner of the LNG storage facility, cannot be reached for comment."

Chapter I

THE WILD SPIRIT

A strange vapor was coming from a rift in the rocks on the slope of Mount Parnassus. Ancient Greek shepherds noted that it caused their sheep to behave as if they were possessed. Believing it to be the breath of Apollo, the god of light, purity, and prophecy, worshipers erected the temple of Delphi over the spot. An oracle was seated on a tripod over the fumes and, in a gas-induced altered state of consciousness, uttered weird syllables. Temple priests "interpreted" the babbling and for centuries before the birth of Christ pilgrims came to seek advice and hear the predictions thought to be inspired by Apollo.

The fire god got the credit for the mysterious vapor at Baku, Russia, long before any sensible use was made of the great petroleum deposits underlying that region. The temple built over that emission attracted devotees from as far away as Persia and India. In recent times, when the temple was demolished, workmen found that a secret pipe had been installed to the altar, thus keeping the "eternal flame" alive.

Three thousand years ago the Chinese used natural gas to hasten the evaporation of brine into salt, sending the gas through bamboo pipes.

The first gas pipeline laid in New York consisted of a ¾-inch-diameter wooden tube.

The word "gas" may have come from the Dutch word

"*geest*," which means spirit. But this "wild spirit" was considered only a nuisance when oil drilling was begun in earnest. When the great Mary Sudik #1 well broke loose in Oklahoma, 100 million cubic feet of gas were flared off per day until the well was brought under control.

Several noted scientists became interested in cryogenics (from the Greek word meaning *born icy cold*) during the last century and began experimenting with the freezing of various gases. In 1917, G. L. Cabot of Boston received a patent on apparatus for condensing natural gas so that it could be used for cooking. But commercial liquefaction did not take place on any significant scale until the Cleveland Natural Gas Liquefaction plant was built in Ohio. It began operation in January 1941. . . .

And so it went, something like that, the first words I ever wrote about natural gas. I was twelve years old, in the seventh grade, and was required to write a term paper on a topic of my choice. Some of what appeared in that paper was gleaned from encyclopedias and other books we had around the house, but much of what eventually went into it came from conversations I had heard at the dinner table as far back as I can remember.

My father was in shipping; my older brother had gone to sea; I knew that I wanted to do the same, so it was only natural that I listened avidly to maritime lore and talk of ships and their exotic cargoes, not the least of which was frozen natural gas, which was just beginning to be carried halfway around the world.

I found it a fascinating subject then. I still do. But with each year I spend at sea and the more I hear about liquefied natural gas, the more my apprehension grows.

Why should we concern ourselves with LNG today?

The recent move to import liquefied natural gas comes in the wake of the U.S. energy crisis. Gas, in its vapor state, is frequently found with oil, and in some parts of the world—

the Middle East, Africa, Asia, and South America—it is still burned off at the wellhead. But since it is relatively inexpensive, burns with a clean flame and, of all the fossil fuels is least damaging to the environment, in the United States it is considered one of the most valuable fuels and provides us with about one third of all the energy consumed.

To transport gas in its natural vapor state from countries which have little use for it would be economically impractical. But chilling the vapors by a very sophisticated process of heat removal to 260 degrees below zero Fahrenheit, shrinking the gas into a supercold liquid that occupies one six-hundredth as much space (i.e., one tanker takes the place of six hundred), continues to be hailed as a major breakthrough and an innovative answer to the energy crisis.

So great has the enthusiasm for LNG been that its transportation has not been limited to ships: an ever increasing number of tank cars and oversized trucks rumble over our rails and roads loaded with supercold liquid gas. Utilities are building LNG storage tanks in almost every major city in the United States. Some of them are the so-called "peak-shaving" facilities which help meet seasonal demand. Those in seaport cities are being linked up to the docks to receive LNG shipped from abroad.

Dozens of LNG superships are planned or under construction in shipyards all over the world. With a price tag of $200 million per ship, none is privately owned, as merchant ships of the past traditionally were. Huge consortiums have paid for their construction through a maze of charter and financial agreements.

Is all the euphoria justified? Is there no flaw anywhere?

The flaw is inherent in liquefied natural gas.

If a large quantity of LNG ever escapes—and it is by its very nature always fighting to escape—the damage to life and property could be as awesome as that portrayed in the Prologue. Several concerned scientists have made similarly dire predictions of disaster.

Comparatively little was known about the hazards of liquefied natural gas when the world's first large-scale liquefaction plant was built in Cleveland. It began operation on January 29, 1941, and after three years of successful operation a fourth tank was added. It too operated successfully for eight months.

Then on October 20, 1944, disaster struck. One of the four cork-insulated tanks suddenly ruptured, spilling 1.2 million gallons of LNG into the dike surrounding the tank. The liquid overflowed the totally inadequate dike and poured into the streets and sewers. Within the confined spaces of the sewer system, the gas vaporized and exploded with tremendous force.

Gas seeped into basements; ignited by pilot lights on hot water heaters, it blew houses apart. Whole families died as they tried to escape the flaming streets, for at the center of the death zone the temperatures reached an estimated 3,000 degrees.

George E. Condon wrote in the Cleveland *Plain Dealer* on the twentieth anniversary of the catastrophe: "The day, I remember, was a lovely day. There was violence abroad in the world at that time in October of 1944, and perhaps this tended to make one more keenly aware of the good things in life, like a day in autumn.

"This was mid-afternoon and two of us on the *Plain Dealer* staff, Oscar Bergman and myself, stood at the rear entrance of the old *Plain Dealer* building . . . and we talked. What we were really doing, I think, was postponing the inevitable return to our desks.

"And then suddenly, off to the east, there was that column of red that rose higher and higher into the sky, close to the clouds, it seemed, creating a scene of terrible beauty.

"'What's the matter?' asked Oscar, pausing in a story he was telling me.

"'I thought I saw a column of flame in the sky,' I said.

"Oscar looked to the east and there was nothing unnatural in the view at first, but suddenly smoke began to roll."

They rushed upstairs where ". . . the city room already was taking on the edgy atmosphere it always does when a big story hits. City Editor Ed Derthick was hunched at his desk, speaking into a telephone, calling up the troops.

"A number of us sped out the Shoreway and up East 55th Street to the disaster site. All day and all night the victims were carried into hospitals all over the city. . . . The others were carried into the old county morgue. . . . Perhaps this was the worst place of all to be, and it's where I spent the late evening hours and the morning until dawn was near.

"The man behind the morgue desk would hold up a pair of charred spectacles, high in the air, and ask if anybody recognized them. Or it would be some other personal effect, like a ring or a keycase. Somebody would shriek and claw his way through the crowd for a closer look, then would be led downstairs where all the bodies were laid on the floor awaiting identification. . . ."

City firemen had been helpless as flames shot half a mile into the sky, and when the fire had burned itself out, 130 persons were dead, 300 injured, and 14,000 left homeless. Ten industrial plants, eighty dwellings, and two hundred vehicles were seriously damaged. The city sewer system over an area of more than thirty acres was destroyed.

So little was left of the LNG tank that it was impossible to pinpoint the cause of its failure, but inferior quality metal, because of wartime shortages, was blamed for the rupture, and the inadequate dike was responsible for the spill.

The Cleveland plant was closed and never reopened.

Throughout the rest of that decade many engineering studies and designs for the liquefaction, storage, and transportation of LNG were made. R. L. Huntington, an expert on natural gas, observed in 1950:

"Though it may sound fantastic and impractical to many,

the proposed transportation of liquefied methane by tanker from South Texas to the Atlantic seaboard has been given serious consideration."

Despite advancing technology, serious consideration took several years. The Cleveland disaster seemed to have proved that the industry had a tiger by the tail. Nobody was quite sure whether to try to tame it or let it go. Transportation involved problems that were not clearly answerable. Further, economic analysis tended to be so ultraconservative that the over-all venture appeared unattractive.

Continental Oil Company (Conoco), however, became actively interested in the possibilities. A task force organized to carry out a feasibility study concluded that ocean transportation of LNG from gas-surplus countries to gas-deficient countries had attractive potential. The company even went so far as to buy up the world's known supply of balsa wood, then considered to be the best tank insulation. (Better insulating materials were soon found and the company was stuck with more balsa wood than all the model airplane kit manufacturers in the world would ever be able to use.)

Other companies joined forces with Conoco and by spring of 1957 complete designs, specifications, and drawings for the liquefaction plant, tanker, and terminal facilities had been completed. Comprehensive analyses of the potential gas sources and markets were made. At this stage a number of countries, including Japan, Britain, France, Italy, and Germany, had expressed interest in importing LNG.

During 1959–60 the first LNG ship, *Methane Pioneer*, made seven trial voyages. Although the ship encountered very heavy seas on several of the trips, no difficulties were reported. It now appeared that the transportation of LNG was possible.

Liquefying natural gas to take care of the so-called energy crisis has been compared to using methadone to cure heroin addiction. Too often in the process of curing any habit the user will encounter other problems which can be just as troublesome. Although there are tremendous advantages in reducing transportation and storage requirements to one six-hundredth, many problems are inherent in the "cure."

1. *LNG is heavier than air.* It spreads out over any surface like batter on a hot griddle. Before it vaporizes and ignites, it will go down into the ground as far as possible, through gratings and into drainage and sewer pipes as it did in the Cleveland disaster.

2. *Contact with LNG or with materials chilled to its temperature of 260 degrees below zero Fahrenheit will destroy living tissue.* The U. S. Coast Guard cautions: "Care must be taken not to spill cargo on human skin." Larry Everman of the Boston Coast Guard safety office was quoted as saying, "They showed us a movie where they stuck a fireman's rubber boot into LNG and the bottom froze and cracked off when they picked it up. Jee*suz*."

3. *LNG in its vapor state is an asphyxiant.* It dilutes the amount of oxygen in the air below that necessary to maintain life.

4. *An LNG vapor cloud flash-freezes anything or anybody in its path.* Since the cold vapor is denser than air, it hugs the ground, forming a wind-borne plume. Controversy exists in the scientific community over how far the vapor could spread, but one of the most highly respected of the experts, Dr. James Fay, of Massachusetts Institute of Technology, says it could extend twelve miles on the full spread of a loaded ship's contents. Other estimates range to as high as 127 miles.

5. *LNG has a flash-back effect.* Ignition at any point in the vapor cloud races back along the entire length of the cloud to the source of the gas, incinerating everything in its course.

6. *No fire-fighting agents have yet been developed that will extinguish a major LNG fire.*

7. *High winds would occur in such a fire.* It would burn with ferocity, developing a hot spot sufficiently high in tem-

perature to form an intense low-pressure area. As this super-heated air rose out of the low-pressure block, the surrounding cooler air would rush in to take its place. Estimates of wind speeds resulting from a large LNG fire range as high as 1,000 miles per hour.

8. *Cryogenic containers are required.* Any but the most exotic, expensive cryogenic materials will fracture and disintegrate on contact with the supercold liquefied natural gas. A mere cupful of LNG spilled on the deck of a specially built LNG ship has been known to crack and open up her deck.

9. *The phenomenon known as rollover occurs when LNG is pumped into a partially filled tank.* The "recipe" for LNG is not always precisely the same, and the various mixes with differing ages, densities, and temperatures fight to reach their natural levels. The consequent rapid rollover can result in tremendous dynamic forces, producing excessive strains on the container walls.

An Esso LNG tanker at La Spezia, Italy, offloaded her cargo after having waited at the dock for a month in the hot summer sun. A few hours after the warmer liquid had been added, a massive boil-off shot into the sky, 440,000 pounds of gas, self-initiated and uncontrollable. Fortunately the wind blew the vapors away from the city and out to sea.

10. *Flameless explosions can occur when LNG contacts water.* These do not occur frequently, but they are extremely violent. Furthermore, they do not involve combustion and cannot be explained.

The U. S. Coast Guard stated in their 1976 edition of *LNG, Views and Practices:* "Many aspects of LNG spill behavior have been investigated with the overall conclusion that the consequences are severe and major spills must be avoided."

No ship carrying LNG has yet been involved in a major accident, but the collision of the Japanese tanker *Yuyo Maru* and the Liberian-registered freighter *Pacific Ares* gives us some idea of what could happen. The *Yuyo* was carrying LPG (liquefied propane gas), the familiar bottle gas, condensed under pressure and consisting primarily of propane and butane. It is much less dangerous than LNG since, except for its flammability, it does not share the previously listed dangers. No shore storage tanks were involved, but it provided one of the worst disasters in maritime history.

I am indebted to a report published by the Maritime Safety Agency of the Japanese government for most of what follows.

On November 9, 1974, the *Yuyo Maru* was heading into the Bay of Tokyo with its four tanks of LPG and tanks of naphtha, another common fuel, which the vessel had loaded in Saudi Arabia. She hired the *Orion*, a vessel with chemical fire-fighting equipment, to escort her throughout the route. Her speed was about twelve to thirteen knots and visibility was approximately two nautical miles. The master of the *Yuyo* was at the conn, and he ordered the second mate to man the radar and the third mate and junior third mate to keep a lookout.

It appeared that the escort vessel did not notice the *Pacific Ares*, partly because of poor visibility and partly because her attention was engaged in guarding the *Yuyo* from smaller craft. The master of the *Yuyo* sighted the *Pacific* for the first time at about 1:32 P.M. when she was approximately one and a half nautical miles ahead of the starboard bow. It looked as if she was intending to cross the route to port ahead of the *Yuyo*.

The master gave the signal on the whistle to draw the attention of the other vessel. The *Pacific* did not change her course. Anticipating collision, the master of the Japanese tanker ordered an immediate stop of the engines and rang full astern. But at one thirty-seven, just five minutes after first sighting, the bow of the *Pacific*, loaded with steel, crashed at an almost ninety-degree angle into the starboard bow of the *Yuyo*.

One minute before impact, the *Pacific* had stopped her engines, gone full astern, and apparently made a rapid left turn in an attempt to avoid collision. Her speed at that time was thought to have been between four and seven knots. The *Yuyo*'s speed had been about ten knots.

The crash of the *Pacific* bow made a twenty-four-meter horizontal opening in the starboard bow of the *Yuyo*, cutting down to the area below the waterline. Torrents of naphtha gushed into the ocean, catching fire instantly and turning the surrounding area into a sea of flames. The chief mate and four other members of the *Yuyo*'s crew were killed and seven were injured.

The bow of the *Pacific* was badly crushed and her whole hull was instantly enveloped in flames. The fire claimed the life of the master and twenty-seven members of the *Pacific*'s crew.

The commanding officer of the Marine Safety Agency's patrol vessel, which arrived nine minutes after the collision, said:

"Immediately upon receipt of distress communication, we rushed to the scene of the accident. Visibility at the time was poor. Suddenly we saw through the mist a large mass of fire extending as long as several hundred meters. At the moment, we thought there was an explosion. We arrived at the scene and could discern the hull of the *Yuyo* on fire. We could not, however, perceive the hull of the *Pacific*, for the whole neighboring sea area was wrapped in flames. . . ."

All available patrols and fire-fighting vessels were rushed to the disaster area. The chief of the air base mobilized Beechcraft and helicopters under his command. Municipal governments of Tokyo, Yokohama, and Kawasaki sent in six more fire-fighting vessels. The Japanese Maritime Self-Defense Force (MSDF) co-operated by sending four of their escort vessels and one helicopter. Private shipping firms also participated in the rescue efforts by mobilizing twenty-one tugboats and escort boats.

Repeated emergency broadcasting from the time of the collision established danger zone areas around the scene of the accident where no vessels were permitted to enter. The Third Regional Maritime Safety Headquarters took immediate charge of rescue efforts. This organization had been systematized for some time beforehand and personnel had been trained to man the headquarters in anticipation of a large-scale marine accident.

Within forty minutes after the collision, thirty-three survivors were rescued from the *Yuyo Maru*. Some of them had jumped into the sea.

The fire on board the *Pacific* was nearly under control at about 5:00 P.M. the same day. There seemed to be no hope that any member of the crew could have survived. But at about four forty-five the next morning a small light was sighted on the deck. Upon boarding, the rescuers found the second engineer crying feebly for help. It was learned later that he had escaped death by remaining in the control section of the engine room, wearing a smoke mask while the fire was roaring aboard his vessel.

Three of the fire-fighting vessels were equipped with seven nozzles each, which were capable of discharging water at the rate of about twenty-five tons per minute. They also carried a great amount of chemical extinguishants. The continuing flow of flaming naphtha from the *Yuyo* made approach very difficult. Not until the fuel on the sea surface was exhausted could the fire-fighting vessels get close enough to discharge chemical foam on the opening in the starboard bow. The force of the flames was still so fierce that the foam flew about in splashes in all directions and did nothing to reduce the fire.

They gave up trying to fight the direct blaze and moved to the port side in an attempt to cool off the adjacent tanks and prevent further burning and explosions. The force of the fire was so great that it was now considered probable the adjacent holds would blow up. Shortly after 4:00 P.M. the On-Scene Coordinator ordered the fire-fighting squad to suspend fighting the fire on board the *Yuyo* and concentrate their efforts on the *Pacific*.

About half an hour later a powerful explosion occurred in the neighborhood of the *Yuyo*'s starboard #2 oil tank. Flames from the explosion rose to a height of about 600 meters (nearly 2,000 feet). The deflagration caused fire to spread to three of the LPG tanks, as well as to reserve cargo oil tanks.

A northeasterly wind was now carrying the burning ships dangerously close to the cities of Yokosuka and Yokohama. It was decided to tow the ill-fated vessels to the central part of the bay. The problems involved in fastening towing lines to large flaming vessels were enormous, attended with the danger of explosion at any moment. It would have to be done quickly.

Patrol vessels and tugboats approached the port quarter of the *Yuyo* where members of the team boarded from the "monkey steps" and fastened the towing line to the bollard on the aft deck. The only flames discernible now on the other vessel were those that continued to blaze in the cabins. The heat was so intense that no one could go on board, so the towing line was fastened to the anchor hook on the *Pacific*'s starboard bow.

Towing the *Pacific* took nearly four hours. The line to the *Yuyo* broke and had to be repaired, so it was nearly five hours before she was run aground on a shoal in the bay. It was not a moment too soon. Immediately after, there was a violent combustion at her bow area. Flames shot skyward.

It was not immediately decided how to dispose of the *Yuyo*. Since there had been no precedent for a fire aboard a tanker carrying such large quantities of LPG and naphtha, it was difficult to presuppose what might now occur. Nearly seventy-five per cent of the naphtha still remained in the tanks. If there were another big explosion, another serious fire could ensue and there would be considerable pollution of neighboring waters.

During the days that followed, explosive bursts of flame occurred repeatedly, hundreds of them. The fire-fighting vessels were on constant duty, discharging huge quantities of sea water and chemical foam.

The decision was made to tow the *Yuyo* fifty nautical miles from the shore. On November 20, early in the morning, towing was commenced. The fire seemed to be stable until evening when the wind became heavier and the waves higher. At seven forty-two, a great deflagration occurred and fire burned on the surface within a radius of about 250 meters. They were then about ten miles from the shore and the tanker had a port list of fifteen degrees.

Nearly two hours later, huge outbursts of flame began again, one rapidly following another. The whole area of the deck in front of the starboard tanks was enveloped in flames. These outbursts caused the outer plating to tear loose from the deck. The fires were becoming stronger and the listing of the vessel was changing from port to starboard.

It was considered too dangerous to continue towing. The lines were released twenty-three nautical miles from shore. The *Yuyo* was a danger to shipping and must be sunk.

The Maritime Safety Agency provides a terse ending to the story of the *Yuyo Maru*:

27th Nov.,	1:35–4:07	Bombardment by four navy vessels
28th Nov.,	9:04–10:20	Bomb and rocket attack by four air force jets
	11:01–1:15	Torpedo attack by a submarine
	3:15–4:16	Bombardment by four navy vessels
	4:47	*Yuyo* sank

In spite of all the Third Regional Maritime Safety Head-quarters personnel, trained in advance, standing by in a state of preparedness, and in spite of all the fire-fighting equipment which rushed to the scene, the *Yuyo Maru* was still burning fiercely as she went down nineteen days after the collision. It had been calculated earlier that unless sunk she would burn for five months.

As a result of the disaster, the Japanese government has given serious consideration to the banning of all gas ships from Tokyo Bay.

I wish to repeat that the *Yuyo* was not an LNG ship. Of the forty-odd ships carrying LNG around the world, none has ever been involved in a major accident.

But the increasing frequency of near misses I experience every time I go to sea convinces me that it is just a matter of time. Ships today are getting into trouble in unprecedented numbers—at least once every hour of the day and night a ship is exploding, burning, grounding, or colliding. Subsequent chapters will explain why I believe that liquefied natural gas carriers will be subject to even higher accident frequency than conventional oil tankers and why I consider the *Yuyo Maru* to be only a minuscule preview of a catastrophe that will boggle the mind.

I hope I have to eat these words.

Chapter II

THE ELEVENTH HOUR

Rossville, in the southwestern part of Staten Island, used to be a seaside resort. Although now in the midst of a heavy industrial sprawl, it still retains something of its small-town atmosphere. Many of the five hundred families who live there chose this village on the shore of the Arthur Kill because it was a good place to bring up children. Safe.

On June 25, 1970, Distrigas, a subsidiary of the giant Boston-based Cabot Corporation, announced in the *Staten Island Advance* that it would construct nine fourteen-story tanks for the storage of liquefied natural gas on a ninety-seven-acre tract in Rossville. Eventually it erected two. These tanks are the largest in the world and each has a capacity of 37.8 million gallons. Together, they would contain enough liquid gas to fill 9,000 of the largest-size tank trucks, a fleet which would stretch, bumper to bumper, from New York to Philadelphia.

Opposition to the construction of the storage tanks was slow to begin. Few laymen at that time knew anything at all about LNG. Five federal agencies, three state agencies, and eight city agencies had given their blessings. Who could fight City Hall?

The residents of Greenpoint, Brooklyn, on the East River, also felt helpless when confronted with two LNG tanks erected by Brooklyn Union Gas. One Greenpointer said it was like waking up and finding that a lion had moved into your back yard during the night. Another said, concerning the gas company's insistence on safety percentages, "Even if the percentages are on the side of safety, percentages mean it will happen—and if it does, good-by Greenpoint!" Antonio Pelligrino, who has raised ten children at his home just eighty feet from the Brooklyn Union Gas property line, said, "They never contacted us. Those tanks came as a surprise."

This lack of due process was to be the keynote of action of communities on Staten Island.

By the summer of 1972 a few protest meetings were being held. To allay the unreasonable panic of a few trouble-makers, Distrigas representatives invited the public to a safety demonstration in St. Thomas Hall, Pleasant Plains, which is adjacent to Rossville. When asked to drop some LNG into a bucket of water, they at first refused, but after much persistence they were persuaded to put a few drops into a puddle in the parking lot.

The few drops of liquid upon liquid produced a large fireball, presumably embarrassing to the safety demon-strators and terrifying to onlookers. It was at that moment that the idea for BLAST was conceived—Bring Legal Action to Stop the Tanks. These people were not environmentalists as such. They had long since resigned themselves to envi-ronmental pollution. For years they had choked on the densely polluted air; swimming was no longer permitted in the Arthur Kill; fishing was a thing of the past, and more than once they had seen the oil-scummed waters ablaze from shore to shore. But they saw the tanks now as a direct threat to their own lives.

Norton Q. Sloan, then president of Distrigas, insisted that the tanks were safe. "There is no kind of conceivable event that could produce damage to structures outside our prop-erty line or pose a risk to human life."

Charles McDowell, also a Distrigas officer, said, "These are the strongest tanks in the world. We have taken so many safety precautions they could stand the impact of a Boeing 747."

Irving Robbins, a physicist at Richmond College of the City University of New York, calculated that when the Rossville tanks were full the two of them would hold the equivalent energy of seventy-four atomic bombs.

At a New York City Council safety hearing, Texas Eastern Transmission Company (TETCO), the owner of an LNG facility in Bloomfield, Staten Island, called Staten Islanders "hysterical" because of opposition to TETCO's proposal to build eight 10.5-million-gallon naphtha tanks next to their stadium-size LNG tank. The date was February 9, 1973.

The next day the TETCO tank, which had been drained ten months before for repairs, blew up.

A cry of "Fire! Fire!" was heard over the intercom just before eyewitnesses saw the enormous concrete dome lift twenty to thirty feet into the air, then plummet at an angle into the tank, crushing forty workmen to death. Pieces of burned insulation were blown all the way across the Arthur Kill into New Jersey.

Only two men, Jose Lema and Joseph Pecora, escaped. They were working on the scaffold stairway about nine or ten feet below the ceiling when Lema heard a sound he described as a low "woof." Pecora also heard sounds from below and said that when he looked around he saw the Mylar lining billowing out "—like it wanted to fly."

He tapped Lema on the shoulder and said, "Let's go!" and started up the stairway with the other man right behind him. Smoke passed them before they reached one of the eight-foot openings in the dome and made their way to safety. Pecora remembered no sound, but Lema said he heard a loud explosion, followed by debris falling all around him.

Christopher Finan wrote three days later in the *Staten Island Advance*, "From the lip of the tank, you can look down and see the tank roof on the bottom like a gigantic jigsaw puzzle—cracked, but intact. For three days and three nights, sixty-odd hardhats and firemen have worked to break up the puzzle and find the crushed bodies of forty men. Topside, eleven pine boxes sit beside a crane. . . ."

A news photo shows the pine coffins and the bodies on stretchers. The sheets covering the bodies cannot disguise the distortion of limbs crushed out of shape and set into strange postures by rigor mortis and the February cold.

Eight days after the explosion, nineteen bodies of the thirty-five recovered had been identified. Bloodhounds were brought to the bottom of the tank in an attempt to locate the five remaining men.

TETCO denied that any gas could have been trapped inside the tank, but the city Fire Department, the Federal Power Commission, and the U. S. Bureau of Mines later were to testify that small concentrations of the gas were found after the explosion.

TETCO denied that the polyurethane insulation and Mylar cryogenic liner were flammable. The Bureau of Mines scientists built a scale model of the tank and treated its foam insulation with hydrocarbons; when they set it afire it burned in a manner similar to the tank which had been destroyed. The Mylar also was discovered to be flammable.

According to TETCO, rigid safety precautions were in force: all personnel were required to wear cotton clothing only—no wool or synthetics—and cotton slippers were to be worn; all personnel were required to remove metal objects such as wristwatches, bracelets, and rings which could cause a static electric spark; smoking was prohibited and all cigarettes, matches, and lighters were to be left outside the tank; all machines and appliances used within the tank were supposed to be explosion-proof.

Some of the charred bodies were identified through rings, watches, and bracelets. Two cigarette lighters were found and also an almost empty bottle of Hai Karate after shave lotion, a flammable liquid with a high concentration of alcohol. Five vacuum cleaners, not of the explosion-proof type, were found. Police discovered, near one of the bodies, a small-caliber revolver with empty cartridges, presumably discharged by the heat of the fire.

Sabotage was suggested as a possible cause. About two weeks before the blast, two employees had been seen punching nail holes through the Mylar lining and seriously damaging it. Co-workers identified them and they were fired. Later the workmen confessed that they had damaged the lining in order to prolong their high-paying jobs.

Several suits were brought against TETCO. The widow of Edward J. Beck, Theresa, mother of one child and pregnant with another, sued for four million dollars. Wendy Rubin, mother of two children, sued for three million.

Angry members of the Center for United Labor Action appeared in force at the St. George Ferry Terminal two days after the accident and distributed 5,000 leaflets. They protested the location of gas storage tanks on the island's west shore and demanded a halt to further construction of the Distrigas tanks. They aimed their protest at safety conditions which were described, but which they claimed did not exist.

Now it was remembered that back in 1966 government researcher Dr. Michael G. Zabetakis had warned against the use of flammable plastic inside the TETCO tanks. Although the Federal Power Commission had requested his opinion, they had ignored his recommendations and approved the plans and specifications.

On Monday following the TETCO explosion, construction proceeded at an accelerated pace at the Rossville tanks.

In April, Mayor John Lindsay, who had made appropriate statements deploring the TETCO disaster in Bloomfield, turned down a request to stop the construction of the Distrigas tank in Rossville.

That same year, after the June collision of the *Sea Witch* and the *Esso Brussels* at the Verrazano-Narrows Bridge entrance to New York Harbor, killing sixteen, Fire Chief John T. O'Hagan warned the Coast Guard that had either of the two ships involved been carrying LNG the consequences could have been cataclysmic. He reportedly said that the bridge would have gone, along with a good portion of Staten Island and Bay Ridge, Brooklyn.

BLAST had been organized late in 1972 with Gene Cosgriff as president. Acting as BLAST's attorney, Richard Sgarlato filed legal action in Staten Island State Supreme Court on October 9, 1973, naming the New York Board of Standards and Appeals and others. The key issue was whether or not the five-man board held stock that may have influenced their decisions.

These are the significant dates and facts revealed by the grand jury report:

▶ On or about June 3, 1965, a person occupying an executive and administrative position with the Board of Standards and Appeals acquired 50 shares of stock in Brooklyn Union Gas Company. On June 21 of that year Brooklyn Union Gas appealed to the board from a decision of the New York City Fire Department opposing the construction of an LNG tank in Greenpoint, Brooklyn. *On September 14 of that year the board permitted the construction of the tank.*

▶ On or about February 14, 1966, a member of the Board of Standards and Appeals acquired 100 shares of stock in Texas Eastern Cryogenics, owned by Texas Eastern Transmission Corporation (TETCO). On October 19 of that year Texas Eastern Cryogenics appealed to the board from a decision of the commissioner of marine and aviation of New York City denying application for a permit to construct in the Borough of Richmond (at Bloomfield) on the grounds that the tank violated the New York City Administrative Code and that the Fire Department also opposed the tank's construction. *On March 10, 1967, the board authorized the construction of the tank.*

▶ On two separate dates in 1969 and 1971 a person occupying an executive and administrative position with the board acquired ownership of a total of 150 shares in Consolidated Edison Company. That company appealed the decision of the Department of Ports and Terminals and the fire commissioner of New York denying its application to construct a 300,000-barrel storage tank in the city of New York. *On May 27 the board authorized the construction of the tank.*

▶ On or about March 24, 1971, a person occupying an executive and administrative position with the Board of Standards and Appeals acquired 100 shares of stock of Cabot Corporation, of which Distrigas was a subsidiary. On May 6 of that year Distrigas Corporation appealed a decision of the Department of Ports and Terminals and the New York City fire commissioner denying its application for a permit to construct a 900,000-barrel LNG tank in the Borough of Richmond (at Rossville). *On July 29, 1971, the board authorized the construction of the tank.*

Three and one half years were to elapse between the filing of this legal action and the release of the grand jury report. Meanwhile, elections were coming up. The transportation and siting of LNG storage tanks were a hot election issue. All those campaigning for the seat of mayor of New York voiced their opposition to LNG.

Abraham Beame, chief contender, was reported as having said back in June when calling on the FPC and the Municipal Services Administration to make immediate interim reports on their investigation of the TETCO explosion: "If these agencies raise the slightest question about the safety of storing LNG and SNG (synthetic natural gas), about the safety of transporting them into and within the city, and about the safety of building more of these giant storage tanks, I will support pending legislation which would prohibit the transfer of these gases anywhere within the city.

"The prohibition of transporting these gases within the city would effectively render useless the existing storage tanks in Brooklyn and on Staten Island as well as those tanks under construction now. The safety of our people must have top priority, and alternate ways of solving our energy shortage would have to be found.

"I do not blame the people of Staten Island for their opposition. Forty men died four months ago when an empty tank in the Bloomfield section of Staten Island exploded while it was being repaired. So far there has been general agreement that the gas company took every precaution for the prevention of the tragedy which did occur. Therefore, if forty men died when a supposedly safe and empty tank exploded, what are we to expect when tanks are filled and one or another precaution is not taken through either human or mechanical oversight?"

What happened to those brave words? Abraham Beame, elected mayor of New York City, ducked behind a statement that he was "only a layman" and didn't know if he had the power to block the Distrigas project.

Frank J. Pannizzo, retained by the Beame administration as first deputy commissioner of the Department of Ports and Terminals, told City Council leaders that "to prohibit an activity which was previously reviewed and approved would clearly be wrong."

Governor Hugh Carey of New York, who denounced the LNG tanks as "menacing" when campaigning, vetoed anti-LNG legislation after election.

When John T. O'Hagan was chief of the New York City

Fire Department he was vocal in his opposition to LNG installations within the city. As fire commissioner of that city, he said he felt the restrictions were adequate.

A company spokesman for Distrigas said, "We build these tanks, taking all the proper legal steps. Now, if they change the rules in the middle of the game, we're going to be left with our rear hanging out the front door."

Federal Power Commission hearings on the importation and storage of LNG in Rossville began in May 1973. Lacking funds as well as qualified witnesses, the BLAST organization was unable to make much headway until October 1975, when they were able to secure, gratis, the services of Davidlee von Ludwig, an arson and explosions expert. For more than thirty years he had worked for insurance companies, gas distribution firms, and foreign governments, investigating more than 2,000 fires, including the Cleveland LNG disaster and the TETCO explosion at Bloomfield. He freely admitted that he had been criticized by some utility and fuel companies whose energy policies and safety standards he had exposed.

Relevant excerpts from the Court's questions and Von Ludwig's answers follow:

Q: Are you satisfied that the Distrigas installations are scientifically sound, safe, and reasonable for use in the area in which they have been constructed?

A: No.

Q: Will you outline the inadequacies or the misrepresentations to this Court which you witnessed during the open hearings?

A: *First*, inasmuch as all massive cryogenic plants are still entirely empirical and experimental in design . . . there is no legal, moral, or scientific basis for making . . . experiments in locales where the lives and properties of persons unwilling to be so threatened are exposed without their knowledge or consent.

Secondly, in respect to the actual materials engineering concept employed in the two Distrigas tanks, they are just as faulty as the error which precipitated the TETCO failure and resultant tragedy. . . .

There does not exist any valid data for the performance of a "pre-stressed" concrete shell liner, supported by the pre-stressed carbon steel wire wrappings detailed to the Court which has been built into the two extant tanks. In the TETCO tanks the Mylar liner was in direct contact with the liquid natural gas and was backed directly with a foot of foam urethane insulation—this exposed the film to embrittlement, but to a fortunately significant extent protected the concrete outer tank from sudden massive area exposure to direct contact with the LNG when the liner cracked from overload embrittlement. This fortuitous alignment of untested materials, coupled with the use of total embankment enclosure of the tank itself, and the use of buried methane detectors which gave warning of the lining failure, *provided a greater margin of security* than the Distrigas configuration, not less, as alleged. . . .

As the composite Distrigas cement/steel liner chills below minus 35 degrees, the difference between the rate of contraction of the concrete and the greater contraction rate of the steel causes enormous increases in the stress imposed on the steel wire—with no residual ductility to absorb these strains, the wires must snap. . . .

Where the concrete tank in TETCO was prevented (by one-foot-thick slabs of urethane foam) from sudden thermal shock when the Mylar liner failed, the concrete interior support in the Distrigas tank (insulated with perlite granules) would fail rapidly from thermal shock failure. This in turn would flood the space between the outer and inner concrete retainer walls of the tank, and unless means of emptying the tank existed . . . the outer concrete would also fail, thus releasing the liquid contents of the tank with ever increasing velocity as the spread of the liquid caused massive additional thermal shock failure of exposed concrete. As these two tanks are NOT earth-wall enclosed, as was the case with TETCO, the actual hazard potential of the two tanks entirely above ground is far greater than was the case in respect to TETCO.

Third, while the Court was shown a room full of instruments purportedly installed to monitor the safety of the systems . . . the television "safety" monitoring system reduces the terrain being scanned to such a minuscule scale on the TV screen as to have no actual value for safeguarding the property. . . .

Fourth, in respect to the much discussed and displayed water, foam, and dry chemical fire fighting systems, they are in fact window dressing for all the value they would have in the event of a massive failure of any part of the system.

Specifically, water is absolutely useless to control a gas fed fire, let alone a fire being supported by liquid natural gas. Exposed to the LNG, water would simply freeze, consequently the hydrants provided for attaching the "super pumper" (assuming such fire equipment could approach such hydrants in an actual failure) are functionally useless in event of any LNG failure other than minor leaks in pipes, etc.

Fifth, the "safety" of the discharge facilities shown for unloading barges, and presumably ships, is open to many doubts. . . . The barge or ship moored at the unprotected dock . . . would be highly susceptible to collision from other vessels attempting to navigate the known treacherous waters of the Kill. The much vaunted "double hull" construction principle is mandatory because it would be obviously impossible to transport LNG in a single hull . . . but present materials engineering design concepts employed . . . are essentially without merit in so far as strengthening the hull is concerned.

Q: In respect to your first line of objection based upon the construction principles employed, how well known are the materials engineering facts upon which your testimony is based?

A: In the petrochemical industry itself the effects of extreme cold in steel and other materials are very well known. In the May/June, 1975, issue of the *Orange Disc*, published by Gulf Oil for stockholders and others, an article entitled "Drilling in the Frozen Mackenzie Delta" discusses the effects of arctic low temperatures on operating equipment:

"At 35 below, the breakage starts—winches break, shock absorbers split, steel handles snap off in your hand, rubber tires become brittle and crack. At 55 below, Gulf regulations say that planes and other equipment will cease to move except in emergencies."

Now these events transpire in equipment which has specifically been designed for use in the arctic slopes . . . they are there confronted with temperatures only slightly below zero degrees Fahrenheit, in comparison with the far colder cryogenic temperatures involved with the LNG systems.

Q: In further reference to the fire control systems shown to the Court, as well as the commentary offered which claimed that no dangerous or combustible gas mixtures would escape from the confines of the Distrigas installations even in the event of any massive systems failures, how valid are such claims?

A: While the inability of water to control a gas or LNG fire ought to be beyond dispute, representatives and officers of Distrigas attempted to persuade the Court that the various tanks of dry chemicals and the various hydrants installed would be both useful and effective. . . .

In small incidents, if there were no unusual wind conditions, prevention of fire via the chemicals would be useful . . . so long as the fume/air mixture which initially would form, while cold and heavier than air, did not reach a source of ignition. Much would depend upon the size of the LNG spill or leak. No data exists on large scale LNG spills, and extrapolation of the small scale tests which have been conducted are invalid. . . .

The Court was told by Distrigas that "calculations had been made which proved that the dikes around the tanks would contain all of the liquid and fumes until the evaporating gas would rise and dissipate harmlessly high in the air; that no combustible mixture would ever reach the boundary of the plant property." . . . No method of calculation exists which can comprehend the unpredictable variables of wind velocity, wind direction, humidity, rain, snow, ambient temperature of the air, the ground, and the water, all of which would help determine what actually would transpire in the event of any massive failure in the pipes or tanks filled with LNG.

Q: Is it your contention that LNG facilities ought not to be constructed anywhere?

A: No . . . given the fact that more natural gas is still being flared off in Arab countries as well as in South American fields due to lack of markets for the gas in proximity to the fields, then clearly the lesser of evils is to liquefy the gas and transport it where it can be at least burned for productive heat purposes. . . .

Let the Gas/Oil cartel construct LNG facilities in unpopulated coastal areas, far from other shipping facilities and far from populated areas. The Sand Barrens of New Jersey on Delaware Bay is one such area which could be utilized.

Q: In respect to the actual safety of the Distrigas installation, aside from the materials engineering criticism offered the Court, what is your present conclusion?

A: Even if it could be proven that no possibility of massive systems failure exists, and if it could be shown that the various fire control systems would suffice to protect the system from all but total failure, two factors preclude any conclusion other than that the system where it stands is a total threat to the security of all persons and properties close to it.

The two factors which militate against the Distrigas installations or any other similar installation in any harbor used on the east coast of this country are the inevitability of tanker disasters . . . and the totally disregarded aspect of sabotage. . . .

In conclusion, I honestly believe and do solemnly swear to this Court that the Distrigas design is even less sound than that employed by TETCO, that the system is inherently liable to catastrophic failure, that fire controls are ineffective, that no means exist to forestall sabotage, that ship movements are inevitably hazardous, and that the system constitutes a grave threat to the security of all within a mile radius, at least.

At the time this is being written, no final decision has been reached by the Court. But the Distrigas tanks, now owned by Public Service Electric & Gas of New Jersey, remain empty. Few would deny that BLAST is almost totally responsible. In the five years since it was organized it has held a multimillion-dollar company at bay with an expenditure of a little more than $5,000. Gene Cosgriff, BLAST's president, is a fighter.

Groups of concerned citizens have organized in many threatened communities. One of the most active has been CADI (Citizens Against Dangerous Installations) in West Deptford, New Jersey, with Lee Joseph, newspaperman, as its chairman.

PACE (People Acting through Community Effort) was set up in the Providence, Rhode Island, area. They held a protest meeting at Broad Street School when permission was being sought by Algonquin to build two 25,200,000-gallon LNG tanks on land owned by Providence Gas Company. The area is surrounded by oil and gas tanks, and only 1,500 feet away from heavily residential Washington Park.

At the meeting Louis Hampton, president of Providence Gas, said, "Algonquin should be congratulated for facing up to the energy crisis and doing something about it. If we thought the LNG tanks were unsafe we wouldn't build them."

Without a public hearing, one of the huge tanks was erected in 1972. It is now partially filled and used for storage. When Algonquin's application to build a second tank was approved by the Providence Building Board of Review, the Citizens Fact Finding Committee, another protest group in the area, immediately took legal action. With decision still pending on this before the Rhode Island Supreme Court, a duplicate application for permit was filed and approved by Mayor Vincent A. Cianci.

Irate opponents of tank construction protested that this was a betrayal on the part of the mayor, who had promised total opposition to further construction of LNG tanks, as well as a highhanded disregard for protocol on the part of Algonquin—how could they apply for and receive a permit now when the first application was being legally contested?

Mrs. Raymond R. West, chairman of the Fact Finding Committee, told me that when the fracas was aired on TV the offending document was shown. But when Thomas McCormick, who works for the Public Defender's Office and is secretary to the Fact Finding Committee, went to the city clerk's office and asked for permission to make a photocopy, he was refused. This document has now "disappeared" from the files of the city clerk.

Some of the hearings before the Providence City Council have been noisy. At one of them, charges were made that some of those who testified in support of LNG facilities have since shown up in high-paying jobs for LNG companies. At another, worried residents adjacent to the tank sites were jeered by representatives of trade unions looking for construction jobs. "Go live in a fallout shelter!" one of them shouted.

The president of Providence Gas Company says plans for constructing the second tank have been aborted. Algonquin's president said that "aborted" was a "bad choice of words." Mrs. West says, "We don't for a minute believe the plans have been aborted. We will not be lulled."

LNG terminals are in the planning stage for southern California and citizens' groups there are protesting, among

them the San Pedro Environmental Action Committee and
the Los Angeles chapter of the Sierra Club, the largest chapter in this country.

Categorized as "bellyachers standing in the way of progress," they seemed to have lost their fight in December 1976,
when the City Council voted its approval of a huge LNG facility on Terminal Island in Los Angeles Harbor. But the
next night an empty oil tanker, the *Sansinena*, blew up and
broke in half in the harbor within sight of Los Angeles. She
was a sister ship to the *Torrey Canyon*, which had been
wrecked off the British coast and caused the biggest oil spill
of all time.

Nine persons died on the *Sansinena* and fifty were injured, many of them children on pleasure boats in the harbor. The blast rocked Los Angeles, tripped burglar alarms,
knocked out phone service, was heard for forty miles, and
shattered windows in Costa Mesa, twenty-one miles away.
About a thousand Christmas shoppers were evacuated from
a shopping center before the cause of the blast was determined.

The incident was of course in no way connected with
LNG, but it provided a mini-scenario of what might happen.
Gregory Smith, a San Pedro educator, long concerned with
keeping potentially dangerous operations out of Los Angeles
Harbor, said, "The explosion instantly made all the eco-
nomic arguments worthless."

City Council members who had initially approved the
plan now favored a review of the project. A bill that would
prohibit construction of the LNG plant was introduced in
the Assembly a month later and a study commission was appointed.

Tom Quinn, in his early thirties, is chairman of California's Air Resources Board and Special Assistant to Governor
Jerry Brown for Environmental Protection. When it was disclosed that a risk analysis report done for the Los Angeles
area's gas utilities estimated that the worst possible LNG spill

could kill 97,000 people, Quinn said, "Clearly, the first LNG terminal in California should not be built near a lot of people. . . . Why, you'd level the whole harbor. What do you think we're into? Genocide?"

Chapter III

BLAST

From Manhattan you go over the Verrazano-Narrows Bridge, take the Staten Island Expressway, drive south on the West Shore Expressway, and you're at Rossville.

It is a morning in late October 1976 when I get there. Using a bit of imagination, I can believe that during the last century it was a sleepy summer resort. Some of the Victorian houses still stand and I see wooded sections with bridle paths. It could be a fine place for a picnic on a nice day.

Which this is not. A raw wind comes in from the waters of the Arthur Kill, bringing intermittent drizzles of rain. But this should be a good place to start asking questions. The citizens' protest group here, living up to its acronym BLAST, Bring Legal Action to Stop the Tanks, has brought more legal action than any such group in the country and so far has managed to stop the owners from filling the tanks with liquefied natural gas. They have been mothballed for three years and it has been costing a whopping $650,000 per month just for maintenance of those empty tanks.

I am getting my first good look at the tanks as I keep driving and keep driving and can't seem to get past them. Stadium size, fourteen stories tall, biggest in the world. I can believe it.

A garbage truck is moving slowly in front of me, the men

jumping off and back on. I crank down my window and call to one of the men, "Do you happen to know where BLAST headquarters is?"

"You mean the Cosgriffs? Gene and Eddie? They live—"

"I'm looking for their office—"

"It's at their house, man. They run it right from there. You're on Arthur Kill Road now and you go to the Maguire Avenue exit. Stay on Maguire till you hit Amboy Road. Take a left at your first light and you're there. Can't miss it. Tell 'em Vinnie sent you!"

I can't find it. I am on Amboy Road now and not seeing any place that looks likely, for I have read somewhere that the Cosgriffs have fourteen children. They would have to live in a big house. I pull into a filling station and ask my question.

"You're looking at it. Right next door."

I drive a few feet farther and park in front of the small brown house he pointed at. I get out my tape machine and some cassettes and walk through the mud to where the house sits among trees. I still can't believe that this is BLAST headquarters and that the Cosgriffs live here with all those children.

Torn screening encloses the porch, which is crowded with wheeled vehicles—a wagon, a tricycle, some skates, a broken-down baby carriage. A dog barks inside as I knock. A bright-faced youngster opens the door and slams it in my face. From the other side I hear, "He's here, Mommy. Mommy, that man is here."

Edwina Cosgriff opens the door, apologizing about its having been slammed in my face. "Come in, come in."

She has a wide smile, nice teeth, white skin that is a contrast to red hair. She's a Juno of a woman in black pullover, black stretch pants—no Scarlett O'Hara waistline, but it's hard to believe she's had fourteen children. As I go into the living room I can see some of them in the adjoining room, the kitchen. It is a Saturday and they seem to be making sandwiches.

"Gene will be out in a minute."

I sit where she indicates on a couch with a spread thrown over it. Some of the springs seem to have died and I change my position so that the ones that still live won't grab me. I look around.

Vinyl-floored, the room has two chairs, one a rocker, an old duplicating machine, file cabinets, TV, stereo. The room is clean, but spit and polish do not have top priority here. I get up for a closer look at the bookcase, which holds a clutter of books and magazines, good ones. I see piles of papers, pamphlets, a couple of scrapbooks, one of which looks as if it has had jam spilled on it.

A hutch cupboard displays family memorabilia—snapshots, school photos, award ribbons. One of them, blue, is attached to a picture of the Cosgriff family taken at a Staten Island family picnic, Edwina tells me, when they won first prize for having the most children. Four times in a row they won, and one year the award was presented by Governor Nelson Rockefeller.

"We had only thirteen then. And when John came along in 1971 we sent out announcements and although there was no BLAST then we referred to him as the latest little blast in the Cosgriff population explosion. We started BLAST in 1972 and John became a real BLAST baby. I took him to all the rallies. His first sentence was 'Tanks gotta go—tanks gotta go.'" She is laughing, turning to introduce the father of the family as he comes out of the adjoining bedroom.

Gene Cosgriff is about six feet tall, with graying receding dark hair. His calm approaches tiredness, but there is a twinkle in the dark eyes. It is hard to see what color they are; the one uncurtained window in the dark-paneled room doesn't give much light. We shake hands and Gene sits in the rocking chair.

"Have any trouble finding us?"

"Not really, but . . ." And then I say, not wanting to be tactless, but saying it anyhow, "I'll level with you. I've read about you, of course, and I figured that, with this many children and all your LNG involvements, you'd have to live in a big house, be independently wealthy, maybe retired, with a staff—"

They look at each other, smiling. Gene says easily, "Eddie here is my staff, BLAST's unpaid secretary. What you see is everything BLAST has got, right here in this room. I'm not retired. A while back, I lost a good job with a tire company —the company folded—and now I sell commercial detergents on commission when I'm not making speeches, attending hearings, trying to organize rallies. We've got a pretty big membership, about eight hundred, but only about thirty are good workers. We call them the 'George group'—you know, 'Let George do it.'"

"And how do you get money for expenses?"

"Donations. Fund-raising events. We've had a little over a thousand dollars a year since we started."

I raise my voice to be heard over the sudden stampede of feet on the bare floor overhead. "How did it all start?"

"Well, the whole thing was under way before anybody understood very much about what was happening. There had been a routine press release in the *Staten Island Advance* in 1970 that Distrigas would put up the tanks for the storage of liquefied natural gas. We heard quite a bit about what a great thing it would be for the area—provide employment, solve the gas shortage. Nobody knew anything much about what the stuff was until we happened to get hold of a study of the theoretical adverse effects of an LNG spill on the Arthur Kill done by Lieutenant Commander Williams of the Coast Guard. We pounded the pavement and got 35,000 signatures empowering BLAST to represent the people opposing the tanks. Then we started writing letters to people we hoped might be willing to help us. Eddie even wrote a letter to Mrs. Nixon."

Edwina is up, looking through one of the files. She says over her shoulder, "One of her secretaries answered my letter—not that I expected a personal letter from Pat herself." She finds the letter and puts on a pair of dark-rimmed glasses and reads: "'It was most thoughtful of you to write to Mrs. Nixon, expressing your concern for the environment . . . we can well understand your distress . . .' And then this person goes on to say that they are referring the letter to officials at the Environmental Protection Agency. A couple of weeks later I receive a letter from Conrad Simon, who is chief of the Air Programs branch of that agency, telling me that it is a local issue and there is no action the federal government can take. I've got that letter here somewhere."

I can smell toast burning out in the kitchen. Edwina looks in that direction for a moment and resumes looking for the letter. An older girl, Lillian, comes through the room eating a sandwich. She's on her way, I gather from the conversation, to her part-time job at a local car service office. The dog follows her out, toenails clicking on the vinyl floor.

Edwina finds the other letter and reads: "'I wish to assure you that the federal government will not permit any facility containing dangerous materials to be operated in Staten Island unless they fully meet federal safety requirements which are designed to make these facilities'—get this—'as safe as the storage tank in your car or the lines carrying natural gas into your home.' This letter is dated February 5, 1973."

The rocking chair creaks (I hear it later on the tape) as Gene leans forward. "February 5. And on February 10 the TETCO tank at Bloomfield, a few miles up the Arthur Kill, blows up. That happened on a Saturday. We thought we had the Rossville tanks stopped for sure. But on Monday they kept right on building, put on double shifts."

I say, "There are LNG tanks in other parts of the city?"

"Yes. At Astoria, Queens, and at Greenpoint in Brooklyn.

At a protest meeting, the chief of the Greenpoint Fire Department—a very honest, forthright guy—said when he was asked what he would expect to happen if he had to try to contain an LNG fire, 'I would expect to die. I would expect all my men to die. I would expect most of the people in the community to die.' I was there and I remember his words very well. We heard that he was reprimanded for that, transferred, and has since been retired."

A boy runs in through the front door and Gene puts out a restraining hand, fingering a shredded shirt. "Hey, what happened?" "Gee, Dad, Eddie and me, see, we—" "Better put on another shirt."

He continues quietly, "The LNG people have produced their experts who say that the tanks here are an 'acceptable risk.' But fortunately for us, there are some scientists who do not agree. Dr. James Fay is one of the top men in the country, a mechanical engineer at M.I.T., and chairman of the Massachusetts Port Authority, and he said right after the TETCO explosion that a disaster at the Rossville tanks would be far worse than the one at Bloomfield. He has called for an experiment in which 10,000 tons of LNG would be spilled—roughly the contents of just one of the cargo tanks on a carrier. According to Professor Fay, there are two reasons why that has not been done. Expense is one—they could probably afford that—but the more important reason is that such an experiment would confirm the catastrophic possibilities of such a spill. And that they cannot afford."

Edwina says, "Mike Kress, a good friend of ours, who is a mathematician and environmental scientist at Richmond College, City University of New York, estimates that if an LNG barge collided off Fifty-ninth Street, Manhattan, and the cloud drifted into New York City, 110,000 persons could be killed. He testified before the FPC and got the gas industry pretty upset."

Gene is nodding. "And then there's Von Ludwig—you know who he is?"

"Yes." I push away the cat who is batting at the tape recorder microphone.

"Accident investigation and prevention have been his whole life. He was testifying at the Staten Island FPC hearings in 1975 with some representatives of the gas company. One of them wrote a note to another—we happened to get hold of it later—and it said, 'Watch out for Von Ludwig—he's a wild man.' Well, he is. He can really fly off the handle and get violent on this subject. They have tried to destroy his credibility. No way. They can't trap him. He has a photographic memory, comes up instantly with names, dates, figures, everything."

My tape runs out and I flip it over. "You've had some politicians who have been helpful too—right?"

"We have. Andrew Stein, state assemblyman, for one. He has introduced strong anti-LNG bills in the Assembly but so far hasn't been able to get any of them passed. Congressman John Murphy is another who has worked very hard. He's an Intervenor in the FPC hearings in Washington."

"What about the attitude of the people—just the ordinary people who live on Staten Island?"

"It varies. Some are fatalistic. Like the mother who said that if the tanks blew up at least the whole family would go together. Another said, well, it would solve the problem of putting the kids through college. Some of the ones who didn't get upset over the prospect of being flash frozen or french fried really got shook up when they found out that sending LNG in here was going to raise the cost of their home insurance policies. The nuns out at Mount Loretto, the missionary orphanage, worry about their kids—five, six hundred of them. They're out on Raritan Bay and they've seen ship after ship grounded and holed there. They know that if one of them should ever be an LNG ship it's good-by orphanage. One of the nuns testified at the FPC hearings."

Edwina: "Sometimes we think we've got everybody behind us. Just before elections we'll have maybe two hundred at a meeting, and then immediately afterward we'll hold a meeting and twenty-five will show up. Sometimes even our paid representatives don't seem to care. At a meeting not long ago I saw one councilman dozing and heard another beginning to snore."

Gene is nodding agreement. "Mayor Beame no longer answers our letters, and when we try to phone, he can't be reached. O'Hagan made all kinds of promises and once he got appointed to be fire commissioner of New York City it was as if he couldn't seem to remember what all the fuss had been about. You could heat the whole city with some of the politicians' hot air. But the gas industry spokesmen are worse. I get sick of hearing how goddamn safe this stuff is. Why, if they're so safe, has the Fire Department said that the next tank they build has got to have a different type of insulation on the floor before it will be approved?"

Edwina: "Tell him about the safety inspector."

"Yes. A safety inspector for Distrigas quit his job and came into the *Staten Island Advance* and dumped, so we were told, a whole armload of stuff on the desk of one of the reporters—memos, documents, safety inspections, you name it. The reporter promised anonymity and ran a couple of stories about it. He quoted the inspector as saying there were thousands of holes in the barrier on the floors of those two tanks, holes made by tools dropped by the workmen. That's how thin those nine per cent nickel steel barriers are."

I get up to retrieve my copy of *Marine Engineering/Log* that one of the boys has borrowed. It is on the big table in the kitchen with bread, jam, peanut butter sitting around. Gene's voice follows me: "We hear so much about safety regulations, but they by-pass them whenever they feel like it. One of the first LNG tankers that came into the Port of Boston was a leaker, so I've been told. They just brought it in and unloaded it as fast as possible. They've got regulations all right, but no punishment for infractions."

"Right." I come back in and sit down again.

"Distrigas put up tanks at Rossville that were seventy times—*seventy times*—the maximum size allowed by city ordinance. But they got a variance. Ever try to get a variance to put in a new bathroom? The TETCO people at Bloomfield had problems with their first and only LNG ship delivery—the level started falling and that clued them into the fact that there was a leak somewhere. That's why they had to take the tank out of service."

Edwina: "Tell him about the lost specifications for the TETCO construction."

"Yes. That. Big thing about it in the paper. They were 'lost' two times, and finally the FBI was supposed to have them—only nobody could ever get a look at them."

Edwina is opening a sliding door in a cabinet. It is crammed with papers, but she knows just where to find what she wants. "Some tapes were lost too. We have this letter from John Lindsay, written when he was mayor of New York, in answer to our inquiry about the safety of the tanks. It says, 'Municipal Services Administrator Milton Musicus, who is chairman of the Mayor's Interdepartmental Committee on Public Utilities, is currently leading a study of the entire LNG storage situation.' He goes on to say that somebody would be in touch with us. Nobody ever got in touch. When I called up about it I was told that the study had been made, recorded on tape, and lost."

Gene gives me a one-sided smile. "That remind you of anything—like Watergate? If they ever do a full-scale investigation of this thing it will make Watergate look like a trickle, a mere drip."

"Have you ever considered getting out of the whole thing, just moving somewhere else?"

"Sure. That's what a lot of people would like to do. You must have noticed all the For Sale signs. John Quinn, who lives closer to the tanks than anyone, has given up trying to sell. It's ruined his wife's health, having to look at those things and worry about them. He has been told that if anything went wrong he and his wife and kids would have about one second to get out—I mean, how fast can you run?"

I said that I had talked to John Quinn. "I think he told me his house is only five hundred feet from the tanks."

"That's about right. Now our house, as the crow flies, is about two and a half miles away. But where would we go? How would we ever find another house with all these kids? You move into a new community with this big a family and you don't get a very warm welcome. When we first moved here twelve years ago, nobody really wanted to know us. Then we had a fire that practically gutted the house. Father John Gordon, our pastor, took us all into the rectory and then everybody rallied around and things started being different. It was Father Gordon who suggested that we start BLAST. We're part of this community now and we care about it. We'll stick it out."

"Have you ever thought that it may be dangerous for you to be fighting the big guns like this? After all, you're costing them a bundle."

"It's certainly crossed my mind."

Action has accelerated now and for a few minutes the place is like a day camp. I don't know which are the Cosgriffs' kids and which are the neighbors'. Edwina gets up calmly to shut the front door, which has come open. "One thing we do here that we think sort of helps protect us is to spread around all the information we get, just Xerox it and get it out as fast as we can. Another thing, whatever we release is not our opinion. In our newsletters, we just quote the experts. And we try to be fair."

Gene, whose eyes have been going back and forth watching the kids, says, "We wish people would be fair with us. We did this program for NBC. They promised to show both sides, equal time. The girl who was producing, writing, and directing the thing spent four or five hours one Saturday afternoon with Von Ludwig. He gave her a lot of information, but she left out all he told her. She spent a full day out here

with us. Later she came back and said, 'Can't you show a little more enthusiasm in front of the camera?'"

Edwina: "What she wanted, I guess, was for us to act like
a couple of emotional hillbillies."

Gene shrugs. "I'm just not like that. I told her that I don't
get emotional. So we didn't give her what she was after. By
the time the film was shown she had practically edited us
out. I happened to be in the hospital recovering from major
surgery when I saw it, and I swear it set me back. The whole
thing was so biased I just couldn't believe it."

"And there was nothing you could do, I suppose."

"I tried. I raised so much hell with NBC that they set up
an appearance for me on the Joe Michaels TV show. He
took issue with everything I said during the interview and
concluded by telling his audience, 'I don't know whether we
clarified anything, but at least you heard a discussion, uninformed though it may have been.' *Uninformed.* Things like
that wear you down after a while."

Edwina is smiling. "But we've won a round or two. Like
the time we blocked the cement mixers with the baby carriages."

"Really?" I check my tape recorder to make sure it's getting this. "When was that?"

"Summer of '73. They were building the tanks and had
twenty-six huge truck mixers of quick-setting cement going
round and round and we blocked 'em, three hundred
women pushing baby carriages back and forth across the
plant entrance gates. The truck drivers went wild, all that
cement setting up. Then one of the big shots came roaring
down to the facility and got into the lead cement truck and
said he was going to drive right through us. And Gene—he
knew one of the cops—yelled, 'Arrest that man—he's driving
that thing without a license!' And the cop dragged him
down from the cab. It was great. We just kept moving back
and forth until all the trucks had to turn around and get out
of there before the cement hardened."

Gene: "Another time we kind of had some laughs was when, just this fall, they sent a representative from Brooklyn Union Gas over to prove how safe LNG is. We were having a meeting at our church parish hall—St. Joseph-St. Thomas—just three doors down. We had a pretty well informed group that night and they kept asking questions and getting no satisfactory answers. This guy was holding a beaker with a tiny amount of LNG in it. There was a visible vapor cloud, very rich, too rich to ignite with this sparking device he kept clicking and saying, 'See, you just can't get this stuff to burn.' Suddenly a five-foot fireball went *whoosh!* It scorched the ceiling and singed the back of his hand. His eyes were like *this*, he was so scared."

Edwina: "This is the second time we've been treated to a demonstration that has backfired."

Gene: "That's right. The first time they poured a few drops into a puddle in the parking lot—liquid on liquid—and this time we saw what happens when vapors come into contact with a source of ignition. All they managed to prove was that here are at least two areas that they don't understand."

His voice is so soft that the sound of his rocking chair makes it hard to hear some of his words on the tapes when I try to listen to them later. I am reminded of a history professor I once had whose dignity and quiet authoritative way of speaking used to have the whole class leaning forward so they wouldn't miss a word.

"Experience. We don't seem to learn much from it. Take that explosion at a nitrate fertilizer plant at Oppau, Germany, back in 1921, when a whole town was lifted right off the face of the earth. Fifteen hundred people vanished, just weren't there any more, and thousands more were injured. The citizens had been repeatedly assured the plant was perfectly safe. Take that explosion at Texas City, twenty years later. It was another one involving nitrate fertilizer, this time on a French freighter, the *Grandcamp*. The thing just exploded in one big thunderclap, disappeared. One thing after another went—another ship, the Monsanto chemical plant, butane tanks. The first blast totally destroyed the Texas City Fire Department."

More kids come through the room. Gene keeps on talking. "Once at a meeting where the president of Brooklyn Union Gas was holding forth about how great it is that modern technology has given us LNG, he draws a space age analogy between the Apollo mission and LNG. I stand up and I say I remember the astronauts, Grissom and Chaffee and White, who died in that flash fire during a routine checkout because somebody goofed, somebody didn't quite know what he was doing. I don't use big words. I don't have big words. I just say what I feel."

The house is quiet now. The rocking chair creaks. "I don't object to the use of liquefied natural gas. I just know that they shouldn't put the tanks in populated areas."

I look at this David and his wife who have stopped the Goliaths and cost them millions of dollars. "So what is going to happen—will you win or will they fill the tanks?"

Edwina: "We're going to keep trying. But so will they. They've got so much money invested . . . I don't know. . . ."

Gene: "I know. We are going to stop them."

Chapter IV

LNG SHIPS: AN EXOTIC NEW BREED

> The old, old sea, as one in tears,
> Comes murmuring with its foamy lips,
> And knocking at the vacant piers,
> Calls for its long-lost multitude of ships.
>
> Thomas Buchanan Read, *Come, Gentle Trembler*.

When I left the Cosgriffs that afternoon I drove slowly past the tanks again and then realized that if I continued north a little farther on the shore-hugging Arthur Kill Road I could probably get a different look at a salvage yard that had intrigued me every time I saw it from aboard ship. It covers twenty-five acres and is immediately adjacent to the new LNG docking facilities. Pilots call it Old Man Witte's Place. It may be the biggest floating junk yard in the world.

My low-slung MG couldn't make it all the way down the narrow potholed track that led from the highway, so I got out and walked to the shack that serves as an office. No one was there. Since it was now late Saturday afternoon I wasn't surprised, but I was disappointed. I wanted to talk to the old man who runs the place, ask him what he did with this catchall conglomeration of rotting tugs, ferry boats, barges, and smaller craft of every description. Some had been burned, some wrecked, others had simply worn out past usefulness.

From a mariner's standpoint, this place is a menace. Many of the still floating hulks extend well out into the Kill. We always slow down before we get there, knowing that we could easily suck them loose from their rotting lines. The pilots say that Old Man Witte has been known to sue when passing ships have damaged any of his "fleet." Supposedly, he sells parts, but the jagged peninsula grows rather than diminishes, extending a little farther each year into a channel that is already too narrow for safety. Why the Coast Guard continues to allow such a navigational hazard is a mystery, especially now, considering the close proximity of the LNG docks.

Gulls cried, foghorns droned, and the tolling gongs of the channel marker buoys kept time to the rocking of the waves. It was now almost five o'clock. The drizzle of rain contributed to the gathering gloom.

A black river rat darted ahead of me as I jumped out onto a half-sunken tug which listed, its stern awash with smelly polluted water. Crunching through the thick rust scales of the deck, I made my way forward where I climbed the ladder to the pilothouse, my hand on the rickety rail. It was not metal, as on vessels constructed today; this relic was teak, deeply textured and bleached from decades of exposure.

The door of the house creaked a little, hanging by a single rusted hinge. The windows had been smashed; the ship's wheel was missing, but the rudder indicator of tarnished brass was still in place, its arrows pointing left and right. The compass was gone; the binnacle remained. I put my hand up and pulled the whistle cord, knowing it would be silent.

A ship went by and as the water was displaced the whole mass of dead ships groaned to life, shifting, straining, clanking.

I tried to see if I could find a name, but it had been obliterated by barnacles and rust. Traces of green paint here and there made me sure this was an old Dalzell tug, a counterpart of the one—or it could be the very same one—that my father had arranged for me to work on when I was fourteen years old. I remembered my excitement then, but now at twice that age I saw the maritime world through different eyes.

The light was fading fast, but I prowled a little longer, finding an old lightship with flakes of her traditional bright red paint, two ferries that used to provide the nickel ride between lower Manhattan and Staten Island, and a dismasted wooden-hulled steamer. She differed very little from the vessels that plied the seas centuries ago.

Alan Villiers, president of the Society for Nautical Research, said in his book *Posted Missing* that seafaring has changed more in the past ten years than in the previous ten thousand.

As I drove back to the city I couldn't get it out of my head that everything was escalating too fast. I knew that by this time next year I could be captain on an LNGC (liquefied natural gas carrier). The prospect made me uneasy.

Is the LNG carrier really so different?

Yes. It involves differences more dramatic than the switch from the paddle wheel to the propeller. It's like an instant change-over from sail to steam. It's comparable to an overnight shift from wooden hulls to welded steel construction.

The world's first oil tanker, the 300-foot steam-powered *Glückauf* (meaning "good luck" in German), was launched in 1886. Today's conventional tankers are essentially the same as this prototype: still steam powered and still single screw, although they have grown over the past ninety years to four times the *Glückauf*'s length. She ran for seven years before skeptics accepted her as the shape of things to come.

In 1969 only four small LNG ships were in existence, but the looming energy crisis with its irresistible profit potential made eyes light up with dollar signs around the world.

The *Methane Princess,* then the largest LNG ship afloat, had a capacity of 27,400 cubic meters. (A cubic meter equals approximately 264 gallons.) The *LNG Challenger,* of a different design, with a capacity of 87,600 cubic meters, took the lead in 1974, followed quickly by the 120,000-cubic-meter *Ben Franklin,* of still another design. Although the *Ben Franklin* was still unproven, nine months later there were forty-three even larger LNGs of various designs on order.

Lloyd's Register announced in September 1976 that "approval in concept" had been granted for a 330,000-cubic-meter vessel, more than twelve times larger than the *Methane Princess.*

Ton for ton, LNG ships are about four times as costly to build and operate as a conventional supertanker. Today's standard size costs in the area of $200 million to build and around $100,000 per day to operate. One cargo valve can cost as much as $40,000. Although these vessels are privately owned, the United States Government subsidizes both their construction and operating costs.

The appearance of an LNG ship is totally functional, distinctively different, one type having five domed-top giant tanks, like igloos, protruding high above the main deck line. Impersonal adherence to construction requirements sets the mood as blueprints are translated into steel with no attention to aesthetic appeal. Consistent with utilitarian design, some of the newer ones bear names like *LNG 42.* The traditional "she" or "her" is becoming a thing of the past.

Throughout 1976 and into 1977, I had occasion to observe the construction of three 125,000-cubic-meter sister ships at Avondale Shipyard in New Orleans. Their decks are jammed with pipes and valves so that catwalk footbridges are required—a far cry from the flat, relatively uncluttered decks of oil tankers where deck hands often ride bicycles as they go about their daily routines.

At the General Dynamics yard in Quincy, Massachusetts,

the largest crane in the western hemisphere, the "Goliath" with a lift capacity of 1,200 tons, is being used to lift the spherical storage tanks into the hulls and place the deck-houses aboard the dozen LNGCs under construction. The 82-foot depth of the hull, measured from keel to weather deck, plus the eight-story high-rise living accommodation aft, require that the radar mast be hydraulically collapsible to pass under even major bridges such as the Greater New Orleans Bridge.

The only part of an LNG ship that seems to be made to human scale lies within the confines of the actual living quarters. To duck through a watertight door leading out on deck is to enter the macro-world of Gulliver. Some wrenches are movable only by a crane. Valves are so large that, instead of taps, wheels two feet in diameter take six strong arms to open and close.

The increasing number of LNG tank designs coming onto the market fall into two basic categories: an independent self-supporting spherical tank, and the membrane tank that conforms to the shape of the vessel's hull. Almost all of the ships carry five tanks.

Whereas the inside of an oil tanker's cargo hold is criss-crossed with bulkheads, swash plates, web frames, and the massive keelson, which is the backbone of the ship, a viewer inside the spherical LNG tank gets the impression of being within a vast mirror-surfaced ball, 120 feet in diameter, or, in the case of the membrane tank, a silo with almost nothing to obstruct the view. The tanks have no visible framing members inside; they have only a ladder, a cargo pump on the bottom of the tank, and one cargo pipe running from top to bottom.

A conventional oil tanker has only a single "skin" of steel plates to stand between the sea water and the oil cargo within. The LNG hulls are single also, and when the manufacturers speak of double hulls they include the multi-layered cryogenic tank walls which are built to hold the cold inside and not to protect from penetration without. Since these are thick, this might suggest safety.

Not so. The deep hull penetration caused by the ramming of the *Global Glory* LNG vessel by the tanker *Bonnie Smathers,* as envisioned in the early pages of this book, is entirely possible. I would remind you also that the *Pacific Ares* was traveling between four to seven knots when she collided with the gas carrier *Yuyo Maru,* and even that slow speed was sufficient to tear deep within her cargo tanks.

During my second year at Kings Point—my "sea year"—I was lucky enough to be assigned for work study to the *Savannah,* at that time the only nuclear merchant ship in the world. Foreseeing the possibility of grounding on rocks, impact with another vessel or an iceberg, the design engineers had been determined to make it all but impossible for any force to penetrate the protective barrier surrounding the nuclear reactor. It was sealed off by massive layers of steel, polyethylene, and lead, plus more than a thousand tons of concrete. Any colliding force would have to penetrate more than seventeen feet of material designed for resilience as well as strength.

When Davidlee von Ludwig testified as expert witness before the Federal Power Commission hearings he said, "If the vessel contacting the side of an LNG barge or ocean transport ship moved with sufficient momentum to penetrate the exterior steel hull, the insulation and inner aluminum alloy liner would afford NO EFFECTIVE FURTHER IMPACT PENETRATION RESISTANCE." (The emphasis is as published in the court transcription.)

The tanks themselves are, to put it simply, the largest portable thermos bottles in the world. All designs use massive amounts of insulation to maintain the supercold and thus hold evaporation of the liquefied gas to a minimum. Many insulating materials are being promoted: balsa wood, mineral wool, silicon-treated granular perlite, and numerous types of foam.

No less varied are the tank building materials which include Invar steel, plywood, nine per cent nickel steel, fiberglass, aluminum foil, and cement. The *Marine Engineering/Log* for September 1976 reports that innumerable other developments are being pursued, particularly in Japan. LNG-related equipment is a new lucrative market and manufacturers internationally are rushing their products to the street. Each advertisement is filled with superlatives. Every manufacturer who hopes for a piece of the LNG pie says his is best. Can they all be right?

The LNGCs are the only ships that actually tap into their cargo tanks for fuel. Since no insulation is a hundred per cent effective, a small amount of the liquid gas is always warming and returning to its natural vapor state. As it vaporizes, pressure builds up inside the tank which ultimately would result in tank rupture and failure unless overpressures were relieved.

But rather than vent it to the atmosphere, an energy-wise solution has been reached by the development of gas-fired boilers which utilize this otherwise wasted energy. There is enough cargo in a standard-size LNG carrier to provide fuel for thirteen months' continuous cruising, making nine circumnavigations of the earth.

Noël Mostert in his book *Supership* says, "Probably no class of ship since the age of steam began has been more systematically prone to breakdown than supertankers. Nor has any class of vessel, whether in the age of sail or steam, been less able in the face of disaster to make do, mend itself a little, and perhaps go to where it's going through patchwork or improvisation."

Once during a shipboard emergency our gyrocompass failed. Rushing into the gyro room, I found the unit to be factory sealed so that no one but an electronics expert would even know how to get a look inside. I finally found the service manual—which was written entirely in German.

Given the greater complexity and ultrasophistication used in the construction of an LNG carrier, no one aboard is capable of handling any but the most rudimentary repair jobs. There are thousands of parts that not one of the crew understands. Equipment manufacturers from all over the world are involved in building an LNG carrier. A typical navigating bridge might be equipped with a Swedish depth finder, English radars, German compass, U.S. steering gear, and a Japanese electronic navigation system. Moreover, few shipyards in the world possess the human or physical resources necessary to make repairs to such technologically advanced vessels.

Even if the shipyard has the expertise to handle cryogenic equipment, regulations require that they must obtain separate and very expensive licenses for each of the many systems. Few yards have even one license and they are scattered around the globe. Since the LNGs were built to make long voyages to the gas-rich but undeveloped countries, a crippled ship on, say, the transatlantic North Africa–North America run or the transpacific run between Alaska and Japan could find itself in an almost hopeless situation.

In the event that a disabled carrier arrives at a harbor that does have adequate facilities, a port authority could justifiably refuse entrance if the damaged ship entailed a threat to the port. When a leaking tanker arrived at the port of Boston, it was unloaded as fast as possible and the port authority was much criticized for so doing. But if a ship is in such distress, where can she go?

The crew might abandon her as did most of the crew on Japan's first nuclear-powered ship in the fall of 1974.

According to Reuter's reports, when the *Mutsu* was about to leave Mutsu Bay to begin a series of test runs, hundreds of fishermen tied sixty of their boats to the vessel in a vain attempt to keep her from leaving port. They feared radioactive pollution of their scallop beds.

The ship managed to slip out to sea during a storm the

next day, and very shortly thereafter a leakage was detected. An emergency attempt was made to stop the leaking reactor with old socks and rice balls containing boron, a major element used in sealing off radioactivity, but this was only partly effective.

The ailing ship was warned by local governments and fishermen's unions all over the country to stay away from their ports. The fishermen who had tried to prevent the *Mutsu* from leaving now threatened to put up a blockade of a thousand ships to prevent her return.

Running out of supplies, drifting without power and with only a skeleton crew, the ship was virtually exiled. Only after the government promised restitution of $3.6 million to the fishermen of Mutsu Bay was the vessel allowed, six weeks later, to come back to her home port.

In the hypothetical case of a leaking foreign-flag LNG ship—and until the time of this writing LNG has been transported only by foreign-flag vessels—is it too farfetched to consider the possibility of an international incident if a damaged ship flouted authority, entered the port, and a disastrous explosion occurred?

Those who are responsible for the unleashed expansion of LNG ship construction justify it on the basis of what they call the perfect track record of the LNG vessels in service.

I have before me three closely written sheets gleaned from *Lloyd's Weekly* and compiled by the Tanker Advisory Center in New York on July 3, 1975. Of the twenty-nine LNG ships in service, sixteen reported no casualties, and thirteen reported casualties such as the following:

The *Jules Verne* in the years between 1967 and 1975 reported engine trouble, machinery damage, leaking tanks, which necessitated a three-and-a-half month layup for repair.

The *Polar Alaska* in '71 reported gas leakage from the cargo tank's primary barrier with varied damage. In 1972 she had damage to propeller, tailshaft, ropeguard, stern tube oil seal, with "cause unknown." She had nineteen to twenty fractured deck platings and was laid up for three weeks' repair.

The *Euclides,* a small 4,000-cubic-meter vessel, built in 1971, had in February that same year "Weather damage, strong vibrations, heavy rolling, ⅔ lower rudder missing, cause unknown. Repaired April." In September of that year she had "hull damage, abnormal vibrations, deck fractured, insulation disturbed, most bridge controls out of order." In 1973 the record shows that in January her recooling gas system was damaged and she had defective action on reliquefaction machines, together with helium leakage. That same month she had weather damage to afterpeak bulkhead, deck fittings, with engine room pipes cracked. In December she had cargo pump trouble with four motors burned out, H_2O in cargo suspected, but she continued trading. The next year the cargo blower used for gas freezing, etc., cracked cast-iron casing. In August she sustained contact damage with another vessel at Terneuzen resulting in damage to bulwark plating, roller fairlead. In November she grounded in Le Havre with damage to fourteen bottom plates, propeller, and two shell fractures.

The *Aristotle,* formerly the *Methane Pioneer,* the first vessel to carry LNG, is no longer in existence, and the listing of damages is sufficient explanation of why she was scrapped. During the five years reported she lost her port anchor and had water in the engine room bilges and tunnel so that the vessel had to sail with portable pumps to Trinidad where a broken auxiliary engine crankshaft delayed her eleven days for repair. Her generator broke down at Port de Bouc with damage to crankshaft and bearings. In November of 1966 she had an engine breakdown at sea with "Main engine bearing heavily wiped, resulting in 53 days of repairing." The next year her #4 cylinder liner broke at top flange. Two cargo pumps were subsequently damaged by alleged crew's overfilling with lube oil. The year after that, in September, off Coatzacoalcos River, Mexico, she was stranded for sixty-one hours, sustaining bottom damage and needing tug assistance to refloat. Two months later she lost her rudder in a gale north of the Azores and was towed to Boston for thirty-four days in the repair yard.

No major blowups with loss of life? No damage to harbor cities? Right. But no one can honestly claim that the track record has been perfect.

Captain Richard Simonds of the U. S. Coast Guard sat in the officers' mess on the *Seabulk Challenger* and told us recently, "We've just been plain lucky so far. A major incident must occur. We're just waiting. It's a question of *when*, not *if*."

A marine consultant specializing in LNG gave me some free time not long ago although his consulting fee is upward of $500 per hour. When I asked him how he could account for the lack of major accidents in LNG ships so far he said, "I think the reason they have been operating without major mishap is because there have been so few of them and those few have been subject to extra-special inspection and precaution. However, as they multiply and we find a couple of hundred in operation—as we will very shortly—the odds will take over. People will relax some of the stringent rules and many will become complacent. The moment they let down their guard one of these LNGs, like a sleeping giant, will go KAPOW! Even with their guard up, their good track record can't last much longer. Just as airlines with great safety records lose planes every so often, so it will be with these ships. Time-tested irrefutable statistics make it inevitable. Nobody's luck lasts forever."

With what seems to be an arrogant assumption that storage facilities will be approved and all opposition overcome, billions of dollars are being spent on carriers to import a planned two trillion cubic feet of LNG by 1980. It is expected that two 125,000-cubic-meter tankers will be making deliveries on the east coast every day.

Dr. Edward Teller, the nuclear physicist whose work led to the development of the hydrogen bomb, told a recent state legislative hearing in California that current technical knowledge of possible LNG accidents is on a par with knowledge twenty-five years ago concerning the dangers inherent in nuclear reactors.

He made a cautious recommendation: proceed with LNG shipping, but with a greatly accelerated safety research program.

Chapter V

BEHEMOTH IN THE BROOK

New York, June 2, 1973. The hour after midnight was calm and visibility was good as the freighter *Sea Witch* sailed around the northern end of Staten Island on the Kill Van Kull. She was outbound for Europe; her decks were stacked with vehicles and her containers with books, liquor, and household supplies. The oil tanker *Esso Brussels* lay just ahead, anchored among other ships near the Verrazano-Narrows Bridge.

The harbor pilot, who had just come aboard the *Sea Witch,* ordered a course change to give a wide berth to the oil tanker. Noting that the ship continued to drift slightly to the right, the helmsman reported that the ship was not steering. The pilot ordered hard left. Immediately the captain came, verified the loss of steering, and took the helm.

He tried to transfer from starboard to the port steering system with no effect. Later, someone remembered that the captain exclaimed, "That damn steering gear again!"

For some reason, never determined, the *Sea Witch* continued at full speed ahead, narrowly missing a tug and barge. The pilot blew a series of short rapid blasts on the whistle and then locked the whistle to sound continuously. With the *Sea Witch* heading at full maneuvering speed for the *Esso Brussels,* by now only one and a half ship's length ahead, the pilot ordered the engine full astern.

But the swing continued rapidly to the right. An attempt was made to drop the port anchor. And then the starboard anchor. Both jammed.

Moments after the captain shouted, "Clear the bridge!" the *Sea Witch* rammed the starboard side of the *Esso Brussels*, penetrating about forty feet and rupturing three of the oil cargo tanks.

Thirty-one thousand barrels (1,302,000 gallons) of Nigerian oil caught fire on impact, gushed into the waters of the harbor, and less than a minute later both vessels were ablaze. Carried by the strong ebb current, now locked together in a fiery embrace with much of the spilled oil pocketed between them, the vessels floated under the Verrazano Bridge, engulfing it in shooting flames and stopping bridge traffic.

Miraculously, there was no injury to motorists or damage to vehicles. The burning ships drifted until they grounded at Gravesend.

A spectacular movie taken by the Coast Guard shows the bridge almost obscured by the inferno of flame and smoke, with the outlines of the vessels barely discernible. Impossible though it may seem, thirty-one crew members, trapped in the deckhouse of the *Sea Witch*, succeeded in attracting the attention of rescuers by waving flashlights. One hour after impact, the fire fighters were able to clear a path through the flames and flying debris and rescue the trapped men. Some of them had severe burns, but all survived.

Several other seamen had jumped into the water; they were saved by tugs whose bow fenders were scorched as they maneuvered between the patches of burning oil which extended about two hundred yards around both vessels.

Not all were so lucky. Sixteen lives were lost, among them the captains of both vessels. One body was never found. Pollution of beaches extended for ten miles, and the cargo on the *Sea Witch* continued to burn for two weeks. That vessel, if all goes according to plan, will be refurbished and con-

verted for service as the first coastwise LNG vessel under
the U.S. flag. When I climbed around over her shortly after
the accident, it looked to me as if the stern might be sal-
vageable; the bow was fit only for the scrap heap.

Damage to both vessels was assessed at some twenty-
three million dollars. The Coast Guard, after lengthy inquiry, concluded that the cause of the accident had been the
Sea Witch's loss of steering control, combined with the high
rate of speed.

But we still have, in spite of a proliferating number of
such accidents, no speed limit in any harbor. I have navigated every major seaport in this country and I know it's
just "Put the pedal to the metal." Pilots who are among the
most reckless are referred to as "cowboys," and some of them
perform as if they were bronc busters in a rodeo.

Port authorities can recommend but not control. This is
not to say that an arbitrary limitation to speed would neces-
sarily be in the interest of safety. There are times when a
vessel must have the option to, as we say, hook it up: proceed with all possible speed.

But a re-evaluation is imperative now that we have
LNGCs as large as the VLCCs (Very Large Crude Carriers)
about to enter harbors and inland waterways where none of
these large ships has ever been able to go, and at a speed
that is far in excess of any tankers in the world. They can
travel at twenty-three knots as compared to the fourteen or
fifteen knots for the conventional tanker.

Two factors contribute to the LNGC's increased speed.
First, the finer lines give her an edge over the oil tanker,
which is so cumbersome that the hull form has often been
compared to a shoe box. Second, since a load of liquefied
natural gas is about half the weight of a comparable volume
of oil, she requires about half the oil tanker's draft; consequently there is less skin friction and less water to push
aside.

Depth of water limitations in rivers and harbors have

been the safety blanket that precluded entrance of any but the small "handy-size" shallow-draft tankers. This safety blanket is about to be ripped away by the LNGCs, who draw a mere thirty-six feet of water compared to the oil tankers' sixty-five to seventy feet.

If the Rossville tanks are approved for filling, LNG vessels will be arriving regularly, transiting the twenty-six-mile Arthur Kill. The Army Corps of Engineers published their survey of the Kill in August 1975. No matter how precise, it could not be expected to include every square foot of the channel. It could easily miss a small object like the tip of the rock that tore into the *Olympic Games* late in December 1976, spilling 134,000 gallons of oil into the Delaware River. Or the rock in the Hudson which gouged the *Ethel H.* in February 1977, causing a thirty-mile oil slick. Or the unidentified object which ripped the *Richard C. Sauer*, spilling 273,-000 gallons in the Arthur Kill. I was on the *Challenger* right behind her, and we picked up a bathtub ring that still won't come off.

But that Army Corps of Engineers survey did show a maximum channel depth at mean low water of thirty-five feet. Since the standard-sized LNG ship at no time draws less than thirty-six feet, her bottom would be in contact with the river bed at low tide and immovable in case of an emergency.

The survey concluded that "The channels in the Kill are not deep enough to allow the most economical safe utilization by the deep draft and large ocean vessels that transit these waterways. . . ."

Ivan W. Schmitt, vice-president of the El Paso LNG Company, told the Third Intersociety Conference on Transportation that his company had established waterway approach requirement for their LNG marine terminals at a minimum of five feet below the keel at any stage of the tide.

Ideally, a vessel should have at least half her draft below her keel. At less than this, bottom sniffing begins and vessel control is sharply reduced, as was noted in the projection concerning the *Bonnie Smathers*.

Channel width is another major problem of the Arthur Kill. The Coast Guard considers it too dangerous for an LNG to continue, as other ships do, on up the Kill into Lower New York Harbor and out to sea under the Verrazano-Narrows Bridge. Consequently, regulations require that the LNG ships, with a length of nearly 1,000 feet, be turned around at Rossville where the navigable channel is only 500 feet wide.

U. S. Coast Guard Captain Frank Oliver, retired captain of the Port of New York, told a meeting of the Council of Master Mariners recently that he sees close maneuvering in the turn-around as the major problem for LNG ships operating within New York Harbor.

The El Paso Company, which seems to be especially concerned with safety, has set its criterion for ship maneuvering area at one and a half to two times the length of the ship. So instead of the mere 500-foot channel width at Rossville, something like 1,500 to 2,000 feet would be necessary to satisfy the not unreasonable El Paso LNG requirements.

I recently requested casualty data from the U. S. Coast Guard to see if the situation on the Arthur Kill was really as bad as my many near misses made it seem. I received a prompt reply from Commander W. J. Ecker, chief of the Information and Analysis Staff:

"We have available statistics on vessel casualties for fiscal years 1963–1975. The particular information you request can be supplied. However, you may not be aware of the size and characteristics of the information you have outlined.

"The area between Staten Island and New Jersey is an extremely congested waterway. Casualty records for the period you request will number in the thousands. . . . If, after reviewing the enclosed information, you decide to pursue your original request, we will be glad to furnish you with a printout. The cost will be approximately $50.00."

The commander's letter told me all I needed to know and I saved myself fifty dollars.

Any pilot will tell you that turning a big ship around in a river is tricky business. My friend Captain Sverre Sorensen was piloting the *Edgar M. Queeny* through a routine river turn-around maneuver at Marcus Hook on the Delaware River shortly after midnight, January 31, 1975. The *Queeny* was moving very slowly; the visibility was good, and there were no unusual bends in the river or any mechanical failure. And yet she crashed through the port side of the Liberian-flag crude carrier *Corinthos*, which had just off-loaded about half her cargo of Algerian crude.

A low-order explosion occurred on impact, followed within seconds by several increasingly violent explosions. The *Corinthos* was immediately a mass of flames, and a large section of her starboard deck and hull, estimated to weigh 110 tons, was blown onto the *Queeny*'s bow. The man on anchor watch there was killed as he attempted to run aft.

The *Queeny*, her bow section ablaze, was able to back away from the fiercely burning tanker, which broke and sank, a total loss. Fire fighting continued for twelve hours, and twenty-six persons were dead or missing and presumed dead.

Captain Sorensen, an experienced pilot with an excellent safety record, was cleared of all charges. Crew members on the *Queeny* were cited for performing "in what must be termed a superlative fashion."

The Houston Ship Channel, a fifty-mile man-made ditch, has a density of water traffic that makes it the most dangerous transit in this country, if not the entire world. In places, the navigable channel is only two hundred feet wide and nowhere is it deeper than forty feet. A recent Coast Guard publication said that seventy per cent of all its cargo may be classified as dangerous, and that ninety per cent of the poisons transported by water in the United States go through this channel. Both shores are packed with industrial plants, hundreds of them, the majority of them being refineries, petrochemical plants, and fuel storage facilities.

I have had experiences on that channel that I do not care to repeat.

It's every man for himself. There is no "tower" to control the chaos of these hundreds of vessels moving twenty-four hours a day at unrestricted speeds and in all sorts of weather.

Try to visualize the pandemonium at an international airport if Concordes were mixed on the runways with Piper Cubs, gliders, helicopters, trucks, and baggage carts. On the Houston Ship Channel a supertanker of 90,000 tons may encounter any combination of dredges, tugs, barges, recreational boats, stationary drilling rigs, sailboats, and fishing boats which may be crossing traffic, overtaking vessels, docking or undocking, frequently with unlicensed pilots and captains, not infrequently drunk, or asleep, or watching TV *while at the wheel*.

One winter night on that channel stands out in my memory. I was sailing as first mate and as I stood alone on the tanker bow surrounded by the thick, choking layer of flammable gases being vented from the 13,440,000 gallons of pure jet fuel beneath my feet, I knew that my chances of survival were getting slimmer with each voyage. Recent statistics showing that an astounding eighty-four per cent of seamen were injured every year did nothing to cheer me.

The vessel's eighteen cargo tanks had just been loaded at Shell's refinery with various grades of gasoline and jet fuel that had a street value of nine million dollars. Everything had proceeded on schedule. The pilot was now aboard, a tug alongside, and all hands in the deck and engine departments were at their assigned stations. I was on the fo'c'sle head, the idea being to stand by to drop the anchors if the captain should order it to avert collision.

This practice may have been effective in slowing the headway of a vessel when ships were smaller and weighed less. But vessel deadweight has grown so disproportionately to the strength of the anchor chains that, although standing by the anchors in close quarters remains on the books as an essential to good seamanship, it has little practical value in an emergency.

The Houston Ship Channel at night looks like a carnival. The thousands of dazzling multicolor lights make it almost impossible to see the lighted buoys marking the channel and the running lights of other vessels. We were heading downstream toward Galveston and all went smoothly until we reached Baytown.

As we rounded a sharp bend to the right I looked aft and saw our stern swinging very wide to the outside of the turn, putting the vessel dangerously close to the Exxon loading docks. We had been lucky that there had not been a ship at that berth. I knew from experience that if there had been one there, even if we avoided hitting it, we would surely, at our rate of speed, have broken her lines and sucked her off the dock.

All of this had taken only a few seconds and when I turned around I saw why we had taken the turn so wide. A northbound ship was coming upstream, and for some reason our captain and pilot had evidently agreed to try a starboard-to-starboard meeting and passing.

Article 25 of the *Inland Rules of the Road* states: "In narrow channels every steam vessel shall, when it is safe and practicable, keep to that side of the fairway or midchannel which lies on the starboard side of such vessel."

But here we were on the *port* side of the channel, lined up to pass each other like cars driving down the wrong side of the road. I knew there must be some "safe and practicable" reason for our vessel to be shaped up as she was—perhaps a sunken object, dredge, or other invisible hazard.

Our ship sounded one short blast of her whistle. This signal can mean only one thing: "I intend to meet and pass port to port."

At first I was sure our pilot had sounded the wrong signal. In my judgment, there was no possible way in the quickly diminishing distance between us to realign either vessel for the port-to-port passage indicated. I wished I did not know that a recent Coast Guard study had disclosed that approximately one half the vessels did not even attempt to exchange whistle signals.

The other vessel did not answer.

Our situation was hazardous. The relief I felt when I heard, belatedly, that single blast was short-lived, for only a few seconds later, with the distance between us closing fast, the approaching ship sounded five short blasts of her whistle.

The danger signal. It is used when a vessel is in doubt whether sufficient action is being taken to avert collision. I was sure it was just a matter of time before impact and explosion. Even if a last-ditch order came from the bridge to drop anchor, it would be purely an exercise in futility.

Through my mind flashed the memory of the man on anchor watch who was killed when the *Edgar M. Queeny* rammed the *Corinthos*. To get as far as possible from impact point he had run down the deck aft, away from the bow. Probably the only reason I remained at my bow station was because I wasn't sure my chances of survival were any better dashing six hundred feet aft over the well deck beneath which lay all that deadly jet fuel.

The order to drop anchors never came. There was no way to stop or even slow down with the engines, as the propeller pitch is designed to deliver only forward thrust.

Between one second and the next the vessel's name, *Dolly Turman,* was clearly distinguishable on her sharp bow in big white letters. Just as impact seemed inevitable, the *Dolly Turman* altered her course and passed outside the navigation buoys. As we shot by she hit the mudbank in a twenty-degree roll.

I had only time enough to draw a thankful breath that she was not a ship that carried significant quantities of caustic poisons or flammable gases when I turned to see that another ship, this time a huge bulk carrier, lay in our path only a mile and a half away. My attention had been so completely absorbed by the near miss with the *Dolly Turman* that I had not even considered the possibility of another ship fast on her heels.

Our speed was still the same as when, a moment earlier, we had skimmed past the vessel that now lay on the mudbank. Even if the throttles had been pulled back, a laden 42,000-ton tanker cannot slow down fast enough to do any good when the distance is that short. Worse still, our frantic efforts to get back on our side of the channel where we belonged resulted in our stern having fishtailed so that we were almost crosswise of the waterway. This time there was no mudbank for the bulker to turn to outside the navigable channel.

I looked aft across the vast well deck to the bridge, hoping for some encouraging sign that collision was again to be avoided and saw that our stern was opening a narrow channel barely wide enough for the bulker to pass. In order to accomplish this maneuver, our vessel was now involved in a violent turn to the left, just as an automobile turns into the direction of a skid to break a skidding turn. Just as our stern opened enough for his bow, our bow closed in on his stern so that we narrowly escaped contact at both ends of our ship in rapid succession.

By this time we were in a state of wild uncontrolled turns to the left and right sides of the channel, like a skier wedelning through slalom competition. When I peered ahead into the darkness I saw yet another big loaded northbound ship which appeared to be duplicating our own chaotic left/right course changes, as if it were our mirror image.

Once again it seemed impossible to regain control of our vessel in time. Whether or not the other vessel began to steer the straight and narrow was almost irrelevant since each of us was back and forth across the waterway every few seconds. It seemed too much to hope that these hard rights and lefts made by each vessel might actually be in perfectly choreographed sequence so as to allow passage without contact.

But pass we did with our hulls so close that I was able to see that the face of the man on the other fo'c'sle head was pockmarked.

How long had it taken for all this to happen? I only know that when it was all over and control of our vessel was once again regained I became aware that I stood in a state of shock in pouring rain. God knows when it had started to rain.

I wish to make the point that it was not unusual to encounter three ships in such rapid succession. In fact, we had been exceptionally lucky not to have been surrounded by the usual menagerie of other craft littering the channel, crossing our course, grounded, anchored, dredging, shifting from one dock to another or tied up loading or discharging on either or both sides of the channel.

But an LNG vessel of the size that will soon be entering this channel would have been too big to miss the Exxon docks, the *Dolly Turman*, or either of the two ships that followed her.

When cars became greater in number, traveling at higher speeds, it was possible to widen roads to eight lanes and build clover-leaf intersections. It is not that simple when it comes to altering a river. The Army Corps of Engineers is the government agency charged with the maintenance of inland waterways, and it is a full-time battle for them just to try to keep up with the silting that is constantly making all rivers shallower. A storm or high wind can lessen channel depth by several feet and the effect can last for many days.

The United States has more ports and navigable waterways than any nation in the world. We already handle more than a million vessels a year. No one is satisfied with the conditions of rivers and harbors, but the job is a gargantuan one, with never enough manpower and money. Keeping the mouth of just one river open, the Mississippi, costs the U. S. Corps of Engineers $80,000 per day.

And now the LNGCs are coming. It is little wonder that many citizens in harbors where LNG facilities are planned feel like sitting ducks.

The Federal Power Commission has issued a map designating the sites where enormous receiving terminals have been built, are being built, or are in the planning stage. A guided tour of the United States starting at the top northeast corner of the map would include:

▶ Everett, Massachusetts. Here, in the heart of Boston, a huge LNG facility is now receiving shipments of the liquefied gas from Algeria. Several parks and playgrounds are within the general area of the terminal and shipping lanes, and at a distance of no more than two and a half miles are located Faneuil Hall, the Paul Revere house, the Old North Church, the Old State House, Bunker Hill, and the Charlestown Navy Yard, where the U.S.S. *Constitution* is docked.

▶ Fall River, Massachusetts. Since 1971 the Concerned Citizens of the South End of Fall River, Inc., have legally challenged New England LNG, Inc., three times, succeeding in postponing the implementation of the terminal there.

▶ Providence, Rhode Island. Citizens have blocked for the time being further expansion of an existing LNG receiving and storage facility. This terminal is surrounded by heavy industry and residential development. Although more than forty miles inland, the area is characterized by frequent thick fog. Many fear the possibility of grounding along the rocky glacier-cut shores of Narragansett Bay, or colliding with an abutment of the Newport Bridge under which all ships must pass. Bridge abutments are always a menace to shipping.

▶ Greenpoint, Brooklyn. Giant tanks have received LNG by barge brought up the narrow East River through the turbulent waters of Hell Gate. The tanks themselves are 1.8 miles from downtown Manhattan and the barge traffic is even closer, passing the foot of Broadway. For the time being, LNG barge deliveries have been halted by the New York City Fire Department.

▶ Astoria, Queens. Two miles from midtown Manhattan are the Consolidated Edison storage tanks, now being supplied with LNG by trucks. This facility lies adjacent to La Guardia International Airport.

▶ Logan Township and West Deptford Township, New Jersey. Applications have been made to the Federal Power Commission for major terminals at these two locations, both of which are on the Delaware River. The first is just below Philadelphia, and the second is directly across from the Philadelphia airport.

Lee Joseph, chairman of the West Deptford Citizens Against Dangerous Installations (CADI), told me that one reason for their concern is that Hope Creek nuclear generating plant, under construction on Artificial Island, is midway between the two proposed facilities. Many questions remain unanswered in connection with the possible effect an LNG mishap would have on this nuclear facility.

▶ Cove Point, Maryland. Columbia LNG Corporation purchased more than a thousand acres in this remote area and built an offshore island upon which to locate their LNG ship-berthing platform. It may seem, on the basis of the projects mentioned above, that an LNG facility cannot be in harmony with the civilization it serves, but Columbia seems to have come very close to doing just that.

For environmental and aesthetic reasons, their plans for an above-water trestle were scrapped in favor of a 6,400-foot underwater tunnel between the shore and dock. Whether from goodness of corporate heart or environmentalists' pressure, it is apparent that they have spared no expense, taking out more than a hundred permits, including one called a "depredation" permit allowing broadcast of the sea gull's distress call, frightening defecating birds away from the berthing platform!

I visited this facility in the fall when the leaves were off the trees and saw that there was hardly any "visual pollution" of that scenic area—the towering tank tops were barely noticeable from the new residential development directly across the road from the terminal.

The only unfortunate aspect of the Cove Point site location seems to be its proximity to the Calvert Cliffs nuclear generating plant. I checked the distance on my speedometer and it is a scant four miles. This does make the siting somewhat less than ideal, but so far it is far and away the best-sited LNG facility in this country.

▶ Elba Island, Georgia. The LNG terminal under construction on the Savannah River has a receiving dock that extends out into the river so that every ship calling on Savannah will have to pass close by the LNGs berthed there, both coming and going. But it is away from densely populated areas.

▶ Lake Charles, Louisiana. This facility will require a lengthy transit through tricky congested waterways.

▶ Los Angeles, California. The planned facility on Terminal Island, in the heart of Los Angeles Harbor, is surrounded by industrial, recreational, and residential development. There are 165,000 people living within five miles of the designated area. Aware that citizens have been questioning the wisdom of siting LNG facilities in that earthquake-prone area, I ordered the five-volume *Final Environmental Impact Report* prepared by the Los Angeles Harbor environment staff.

Approximately 150 pages were devoted to endangered marine species such as *Neanthes arenaceodentata* (a worm) and the *Pachygrapses crassipes* (a crab) but I noted only about ten pages that had to do with earthquakes. Worried citizens may not derive much comfort from the promise that "if an earthquake of sufficient magnitude occurs the entire LNG facility will be shut down promptly."

▶ Oxnard and Point Conception, California. These sites lie up the coast, south and north of Santa Barbara. Oxnard has more people, 30,000 within a ten-mile range, but Point Conception, with a population of only 500 within a like area, has scenery that environmentalists are fighting to save.

▶ Newport, Oregon. A ten-million-gallon tank has been built here on landfill. This is a port of entry for LNG from Alaska. Residents are fighting.

▶ Portland, Oregon, and Tacoma, Washington, are designated as major potential sites.

▶ Valdez, Alaska. The proposed huge gas liquefaction facility here provides for 165,000-cubic-meter LNG vessels, the largest yet, entering waters where tides rise and fall as much as forty feet. An Alaskan state survey predicts a high probability of major accidents to tankers navigating the Valdez Narrows.

According to the *American Maritime Officer*, January 1977, the hazards include "freakish 200-knot winds, strong currents, and an outcropping called 'Middle Rock' in the center of the 6,000-foot-wide channel."

Valdez is in one of the most active earthquake zones in the world and was heavily damaged in the quake of 1962.

Chapter VI

DOWN TO THE SEA IN DOLLARS

To be the first vessel to cross the Atlantic powered by steam alone! It was an ambitious goal and the *Great Western*, her hull trussed with iron, sheathed with copper below the waterline, had been specially built to meet such a challenge. She left England bound for New York on April 8, 1838, confident of a place in history.

When she reached New York fifteen days later she found that the *Sirius*, half her size and built to run only between England and Ireland, had beaten her record by four hours. A rival company had schemed to win the record by sneaking the *Sirius* out three days before the larger ship left port, and the master of the vessel had made it happen by burning cabin furniture to keep the boilers at full steam ahead.

Such competition remains the cornerstone of shipping. Beating the deadline keeps that bottom-line figure in the black. And that is what counts in the maritime world.

In the days of sail, at the mercy of unpredictable weather, imprecise charts, and a host of other uncertainties, the fact that ships got there at all seemed something of a miracle. A captain might get to know the ways of ocean currents and learn how to cope with the ever changing direction and intensity of winds, but he could never command those winds to fill his sails and take him where he wanted to go.

Vessel scheduling was born with the advent of steam

power. Aided by improved weather pattern knowledge, greater chart accuracy, more precise navigating techniques, and improved ship construction, shipowners began to concern themselves with *when* their ships would reach their destination, not *if*.

With "deadlines" the key word, shipping became a numbers game as never before. A shipping company's success depended on its ability to adhere closely to a predetermined schedule. Prevoyage calculations, accurate to a fraction of a percentage point, revealed a shipowner's competitive edge—or lack of it—before a vessel ever left the dock.

In a tight market, profit margins can sometimes be carved only from corner cutting.

Daniel K. Ludwig, multibillionaire, probably the richest man in the world, is the world's biggest individual shipowner in terms of tonnage, and the unquestioned king of the corner cutters. Every ship in his fleet (National Bulk Carriers) is stripped of all but the essentials. Even the captain has spartan quarters. Someone who knew his business philosophy once suggested that an appropriate logo for the Ludwig fleet flag might be two hands stretching a rubber dollar bill.

Any seaman can recognize a Ludwig tanker at a glance: a totally black hull with deckhouses, masts, and other deck paraphernalia reduced to such a minimum that it looks as if it might have gone under a low bridge at full speed.

A mariner friend with years of experience aboard the National Bulk Carrier fleet told me that in the early days Ludwig frequently visited the ships he owned. He remembered asking D.K. where he wanted a piece of steam pipe stowed. "Steam pipe?" roared D.K. "That's the smokestack!" He is reported to have said when asked why his ships didn't have the huge smokestacks, decorative but usually fake, displayed by other lines, "They won't hold oil, will they?"

It may not be true, but I've heard more than once that Ludwig employees used to have a sort of slogan: "Another two inches for D.K.," which meant overloading the tanker beyond her allowable limits and then listing her slightly, just enough to keep her Plimsoll mark above the water's edge when viewed from shore. Even if caught, with the fine then nowhere near as large as the extra revenue gained, economically it made sense to squeeze in a few extra barrels of oil and load deeper by "a couple of extra inches for D.K."

Stavros Livanos, one of the "Golden Greeks," is another tightfisted shipping magnate who visited his ships frequently. I never could quite believe the story about how he would shake hands with every crew member and then immediately fire some of them, explaining that if a man's hands were soft it was a giveaway that he hadn't been working very hard. But a witness has assured me it is indeed true, and that furthermore, when a seaman once ran to escape the handshaking routine, Livanos chased him. Thinking the older man would never follow him, the seaman climbed the mast. But Livanos came right after him. He shook his hand. And then he fired him.

Elimination of non-productive personnel takes place ruthlessly at all levels, even the highest. Money-minded shipowners seek out captains who will be loyal at all costs. These men know that if they are not on management's side all the way they can be replaced easily. To call them bootlickers may sound harsh, but I have sailed with captains to whom only harsh designations apply.

They try to keep structural defects secret from crew members, and even sometimes from the owners, some of whom would prefer not to know. They will put to sea in an unseaworthy condition if it pleases the company; to meet a deadline or even an owner's whim, they will disregard not only the *Rules of the Road* but the most basic dictates of good sense.

A blatant example of which I have personal knowledge took place on a tanker which was driven without mercy through storm and heavy seas. She suffered visible exterior damage, possible hidden injury, endangering not only the ship but all hands aboard. Short of mutiny, a crew is helpless, but this crew was enraged. They knew that the reason for this dangerous haste was a cocktail party the owners had scheduled aboard that ship at her next port of call.

One of the printable accusations about this particular captain is that he supposedly has the company emblem tattooed on his chest.

Will deadlines be allowed to take precedence over everything else when American-flag LNG ships come into the picture? Opinion is divided. Captain Warren Leback, vice-president of the El Paso LNG Company, with nine LNG ships under construction, says no: "The masters will be under no pressure to meet delivery schedules when safety is involved."

Joseph Cuneo, president of Energy Transportation Corporation, and a leading figure in LNG shipping, says yes: "The need to keep [them] moving and to avoid unscheduled down time is far more severe than in virtually any kind of shipping." With the meter running at around $100,000 per day, Cuneo's statement would seem to be well founded.

In a good tanker market, corner cutting is almost unnecessary. It's hard not to make money. The cyclic, feast-or-famine nature of the business was summed up in a peak market period (1970) by one shipowner who said, "These days, no one in our business is on the breadline, and some owners are really gorging themselves. And why not? In this business, when the food is on the table you eat as much as your stomach can hold. Tomorrow, you may be able to see your ribs."

From the mid-sixties through late 1973 those who gorged themselves were:

▶ Hilmar Reksten, Norwegian shipowner. He fixed a one-year contract with British Petroleum worth $81 million. His estimated profit was $60 million.

▶ Aristotle Onassis. He took early delivery on a new 200,000-ton tanker that was to go on long-term charter. Rather than leave the ship idle for two months, he leased it to Shell Oil for one voyage from the Persian Gulf to Europe. His net profit from just that one decision was over $4 million.

▶ John Theodoracopulos. He bought an Exxon tanker, small and old, for two million dollars. By the end of the first year he had repaid Exxon in full and netted for himself three million.

Ironically, during this unprecedented period of feasting the move was begun toward larger tankers: superprofits were not enough; tanker owners everywhere wanted more.

The realization that 10,000 tons of oil could be transported almost as cheaply as 100,000 tons—or 200,000 tons, or even 300,000 tons—vastly multiplying the tanker's owner's profit, was irresistible. One huge ship with only a standard-size crew and a single engine could now carry what had required a fleet of ships only a few years earlier.

This stampede toward larger tankers began in the early sixties. Easy financing prompted wildly speculative building. Shipyard order books bulged for years ahead. But the result of it all was that the market became grossly overtonnaged. There were simply too many ships and not enough cargo to go around.

Then in late 1973 the Arabs lowered the boom with their oil embargo and hundreds of tankers went into layup.

The idea that bigger is better was nothing new. As far back as 1852, Isambard Kingdom Brunel explored the economies of ship size in England. His *Great Western* was a success and so was his *Great Britain*. Then he designed the *Great Eastern*, five times larger than the size of any vessel ever built. Brunel's dream was to monopolize an entire trade route with only one vessel. Her construction took five years.

On what should have been the gala occasion of her launching, she crushed the launching gear when they tried to slide her down the ways into the Thames, and was so heavy a windlass broke, flinging workmen into the air. It took all winter to launch her and by then Brunel and his company were broke.

Undaunted, Brunel formed a new company and rebought his *Great Eastern.* Toward the end of the year when she finally began her sea trials, an explosion occurred which killed five crewmen. Brunel himself dropped dead when he heard the news.

Despite everything, this freak began transatlantic trading. In a storm en route to New York, her paddle wheels were smashed, her rudder ripped off. Two cows were flung through a skylight and into the beautifully appointed ladies' salon. East of New York, in Long Island Sound, a reef still bears her name. Area yachtsmen know this hazard well, and when I first began ocean racing at the U. S. Merchant Marine Academy at Kings Point, they were quick to warn me of Great Eastern Rocks which ripped the hull and grounded that luckless lumbering ship.

Thereafter, gutted and stripped of her luxurious appointments, the *Great Eastern* performed humbly as a transoceanic cable layer until she met her fate at the wrecker's yard. When she was broken up, no easy task, wreckers found the skeleton of a riveter who, many believed, had put a curse on her from the beginning.

Despite this failure, the builder of the *Great Eastern* had a great idea. The *raison d'être* was as sound as it is today: bigger *is* cheaper. Shipbuilding and operating costs do not rise in proportion to vessel size. The cost of propulsion machinery is about the same. Crew requirements are the same, no matter how much larger the vessel's size; in some cases, because of automation, the size of the crew actually shrinks.

Economy of larger size, however, was not enough. It did

not afford the attractiveness offered by the flag-of-convenience system. This arrangement is undoubtedly the most tempting incentive yet devised under the law for the prosperity of shipowners worldwide.

Liberia, a small West African nation without even a natural harbor, has amassed the largest and fastest-growing merchant fleet in the world. Panama comes next and a dozen other nations follow.

How can this be? It's very simple: officers do not have to meet the more demanding license requirements of traditional maritime nations such as Great Britain and the United States, and thus do not command the higher wages of these nations. Less stringent safety requirements and infrequent hull and equipment inspections provide scant incentive for costly maintenance. The fact that Liberian authorities accept the licenses of most other maritime nations and virtually none of these will accept the license issued by Liberia tells a lot.

For that small African nation, this system is a piece of cake. With a minimum of expense and trouble, Liberia receives sums that add up to eight per cent of her gross national product. Small wonder that other nations are lined up trying to get into the action.

When I was studying for my second mate's exam I lived at the Seaman's Church Institute on lower Manhattan. Sometimes for a break I used to wander up Park Avenue to the Chock Full o' Nuts shop on the corner, climb one flight of stairs, and shoot the breeze with a friend who worked at the licensing department of Liberian Corporation Services. It wasn't a very large office; there were perhaps a dozen employees, none of whom was any more Liberian than I am.

One day I happened to mention to my friend that the night before when I was studying in the library I had been approached by someone concerning the purchase of a Liberian license—a captain's, no less. He shrugged. Sure, black market. It happened all the time.

Jesse Calhoun, Marine Engineers Beneficial Association leader, has been quoted as saying, "You could be throwing coconuts out of a tree last week and be the master of a vessel this week." I'm not sure which came first—those words or the color photograph that hangs over the desk of the receptionist in the Liberian-flag shipping headquarters: a chimpanzee decked out in captain's dress whites and gold braid has one hairy hand raised in salute. The Spanish caption translates "Salute from the Captain."

This exaggeration captures the unhappiness of all maritime unions with the flag-of-convenience system. It takes the bread right out of their mouths.

By using the Liberian flag, a shipowner can hire any nationality he chooses at enormous savings. According to figures supplied by Philip J. Loree, director of the Federation of American Controlled Shipping, annual wages for a crew of thirty-two U.S.-licensed seamen would run to an estimated $1.7 million per year. An Italian crew would cost something like $600,000 per year. It would be less than that for the British, Spanish, and Greek, and all the way down to the "Brand X" crew, made up of various nationalities, which could be had for about $200,000.

All this money saving results in profits, none of which are subject to U.S. corporate income taxes. Nor does the company that forms the Liberian subsidiary have to pay taxes in that country.

One third of the Liberian-flag fleet is American owned. Shipowners, like Gulf Oil, Mobil, and other large oil companies, provide strong support for the flag-of-convenience program. Exxon, for instance, since 1969 has registered nearly all new shipping under the Liberian flag.

Registering your corporation under flags of convenience has still another attraction: anonymity. This can be a real advantage if your Liberian-registered ship ends up like the *Argo Merchant* in December 1976, aground off Nantucket Island, dumping 7.3 million gallons of oil before sinking.

I read—as I suppose the whole world did—about the *Argo Merchant*'s track record: the skipped inspections, the spills, the fines, the heavy rusting everywhere, the faulty gyrocompass, the outdated pilot chart, the unsanitary living quarters, and I thought—as I am sure other mariners did—So what else is new? We have seen similar conditions on so many ships that it is surprising so much is made of it.

The Organization for Economic Cooperation and Development has estimated the accident rate of flag-of-convenience ships at four times that of vessels registered by traditional maritime nations.

According to Arthur McKenzie, of the Tanker Advisory Center in New York, preliminary figures for 1976 show that a record nineteen tankers were lost. Eleven of them were Liberian, two Cypriot, four Greek, one Spanish, and one East German.

Liberian shipowners say they are not ashamed of their safety record, claiming that their high number of losses is only because they have the largest fleet.

Guy Maitland, executive director of the Liberian Shipping Council, backs this up, adding that some of the most modern, best-equipped, and biggest tankers are in the Liberian fleet.

Robert J. Blackwell, who is the U. S. Maritime Administrator, told the Senate Commerce Committee in January 1977, "We have never seen that fleet." But this did not mean that he was calling Maitland a liar, only that the newest Liberian tankers are too big, drawing too much water for our relatively shallow harbors; consequently, we see only the older, smaller ones.

Some of the older, smaller ships we see around are rented (chartered) by major oil companies, and these are apt to be registered under flags of convenience. These companies operate their own ships to the highest standards of safety, and yet, when for economic reasons they augment their tonnage with substandard chartered vessels, they are unable to exercise any jurisdiction over these decrepit vessels which limp to and from their docks.

When damage is caused by a flag-of-convenience vessel it is almost always the taxpayer who eventually must pay. The Liberian-flag tanker *Garbis* is a case in point.

On the night of July 17, 1975, she discharged 50,000 gallons of heavy crude over the side, blackening fifty miles of Florida beaches—a gooey mess that cost $400,000 to clean up.

Samples from the slick were taken to the lab and an "oil fingerprint" was arrived at. A search of the shipping records of every tanker in the world came next. Gradually the list of potentially suspect vessels was narrowed to 247. The Coast Guard boarded each one of the vessels—the *Garbis* among them—taking cargo samples. Laboratory testing revealed that the sample from the *Garbis* was the perfect match for the sample of oil taken from the Florida spill. Now the Coast Guard had only to wait for the culprit to call at a U.S. port again.

They waited almost four months before they caught the *Garbis* docking at a Delaware River pier in South Philadelphia. The Coast Guard boarded promptly and arrested the captain.

One for the good guys? Hardly. Not in the world of third-flag shipping. Because the Straits of Florida are international, the federal district attorney decided he had no jurisdiction to bring criminal charges against the master of the Liberian *Garbis*. The vessel was released. The United States has been trying ever since to recover with, at this writing, no luck.

In another instance of the Coast Guard losing all the way around, they arrested the captain of a Liberian-flag vessel that was causing pollution in the bay of Tampa, Florida. While they were holding him in jail, the guilty vessel sailed away. Presumably lacking sufficient evidence to prosecute, the Coast Guard had not only to release the captain but pick up the tab for his plane fare back to Greece.

We have no reason to believe that when flag-of-convenience LNGCs start moving in and out of our waters they will be any more accountable. With international maritime law and U.S. environmental and shipping laws as imprecise as they are on the subject of liability, it is all a vast no man's land.

Even if we believe—as we are told—that owners can at least be sued for the value of the ship and its cargo we ask, *What* ship if it's a total loss? *What* cargo? *What* owners if anonymity is guaranteed?

Shipowners have defended the flag-of-convenience system in a number of ways, one being that in the event of war the shipping would be readily available if under the flag of a nation friendly to our interests. This reasoning seems to have flunked the test during the Yom Kippur War when, contrary to U.S. policy, the Liberian flag respected the boycott of Israel imposed by the Arab states.

Another argument put forth by the shipowners is that their ships are under flags of *necessity* rather than convenience, claiming it is not for larger profits but necessary if they are to have any profits at all and be competitive in international trade.

The public cares little whether or not a shipowner enjoys a competitive edge. Most people are concerned primarily with the possibly adverse effects that the so-called runaway-flag vessels will have on them and their environment. A defective tanker, leaking oil with its threat of long-term destruction to marine life, has been alarming enough, but the possibility of LNG being carried in any vessel that is not as safe as the law can make it carries a threat to human life that is totally unacceptable.

Many LNG ships now sail under flags of convenience. The *Arctic Tokyo* and *Polar Alaska,* owned by Phillips Petroleum, fly the flag of Liberia and both trade between Alaska and Japan, exporting liquefied natural gas to the Far East. The *Euclides*, trading between United States and Algeria, is also Liberian, as is the mammoth *Paul Kayser*, owned by El Paso Marine Company. There are others.

Perhaps the most disturbing aspect of the flag-of-convenience situation, particularly with regard to LNG transportation, is in the "hand-me-down" ships, vessels whose best years are behind them. The cost of keeping their aging hulls sound has become prohibitive for the responsible shipowner. And so, in an honest effort to keep his fleet as safe and up to date as possible, he will dispose of spent tonnage.

This ship finds its way into the hands of a here-today-gone-tomorrow shipowner who is in the junk box market only because he cannot afford newer tonnage and needed maintenance. For example, the twenty-three-year-old *Argo Merchant* was on her fourth owner when she grounded.

Captain Warren C. Leback, vice-president of El Paso LNG Company, told the New York chapter of the U. S. Merchant Marine Academy Alumni Association, "There is no room for the marginal or conventional type of shipowner in the transportation of LNG."

Such a statement is more of a hope than a reality.

In its final days the Ford administration approved the largest federal handout to a single corporation in the history of the United States, a $730-million construction loan guarantee to Burmah Oil, a British corporation presently having huge LNG ships built at the General Dynamics yard in Quincy, Massachusetts.

The British government, two years previously, had saved Burmah Oil by guaranteeing its debts to international banks. Loss of the LNG contracts would have caused the collapse of the company and default on its loans, over $500 million, with Great Britain responsible for repayment.

So it appears that at least one "marginal" shipowner is already involved in the LNG business on rather a grand scale.

In order to understand the seemingly sudden emergence of so many LNG ships, a bit of backtracking may be in order. Mention has already been made of the fact that by the beginning of 1974 the bottom had fallen out of the tanker market. It had been so good for so long that far too many ships had been built and ordered. The Arab oil embargo had reduced the amount of oil available for transport. Prices had quadrupled. The U.S. economy seemed to be slipping into the worst depression since the thirties. These factors and others combined to create the worst tanker market in the annals of maritime history.

Hundreds of tankers, mostly VLCCs (Very Large Crude Carriers) went into layup. Those that were lucky enough to keep running did so at reduced speed. Shipyard order books dwindled with cancellation after cancellation. Many shipping companies, especially those building on speculation, went bankrupt, leaving unfinished orphaned behemoths on the building ways.

The best employment one tanker owner could find for one of his idled ships was renting it to the makers of *King Kong,* who used one of the ship's vast tanks as a cage for that mechanically contrived monster. A supertanker reduced to a mere movie prop was an ignominious symbol of the times.

Even some of the Golden Greeks panicked as fortunes evaporated while mortgages and other fixed costs continued with no income to balance the scales.

"I christen thee *Olympic Bravery!*" It was October 1975 when Christina Onassis smashed the traditional bottle of champagne against the bow of the *Bravery's* glistening white hull. This was the first VLCC to be delivered to the Onassis fleet since Christina took over after the death of her father.

Thirty-five miles into her maiden voyage, the 275,000-ton *Bravery* was rolled over and beaten to death against the jagged rocks of the island of Ushant.

Time's headline read: "Maritime Disaster—or is it?"

Consider the circumstances. The *Bravery*'s maiden voyage was to have dead-ended in a Norwegian fiord, joining almost four hundred other idle tankers. I have seen these tankers and it is a depressing sight. For her to have headed there on her maiden voyage was comparable to a debutante leaving her ball and heading for the old ladies' home. Even that was going to be expensive. Her layup costs would be about $20,000 per day, what with interest, amortization, insurance, and maintenance. If this $50-million ship had been offered for sale, a buyer might have been found at $15 million. Hardly more.

Almost immediately it was announced that Lloyd's of London, together with other insurers, would pay the insured price of $50 million to the Onassis group. To date, the loss of the *Olympic Bravery* is the most costly mishap in maritime history.

The Greeks were not the only ones in trouble. Japan, which since the late sixties had been responsible for about half the new tonnage, was backed against the wall with Spain, France, and Sweden, all wondering what they could do to occupy their yards during the bad market.

The LNG ships provided the light at the end of the tunnel.

U.S. shipyards had been uncompetitive on a world scale, but the extreme sophistication of LNG vessels helped give the U.S. yards an early lead since they were better able to compete in the more technical areas of ship construction.

The scramble was on. Yards all over the world were trying to wrap up as much of the business as possible, and the best way to do that was to offer the fastest delivery at the lowest price.

It is this aspect of the situation that worries many of the knowledgeable persons in the industry. It all seems too reminiscent of the way many of the hastily prefabricated tankers of the sixties were flawed. Some of the Japanese ships had cracked soon after they started trading; some sailed with vital parts of their structures not welded at all. Other countries also had been guilty of building too big too fast without waiting for sea experience to test wearing qualities and response to stress.

But the oil tanker escalation was puny by comparison to what is happening with LNGCs.

Speculation about the future of the LNGCs as a species seems to spur the builders on. Other solutions to energy needs may phase them out and nobody can be sure just how long the action will last, but right now they are *hot*.

General Dynamics at Quincy, Massachusetts, contracted a few years ago for a dozen or more LNGCs at a price of $90 million per ship, which now has escalated to a unit price of $200 million. At this writing, not one of those ships has been delivered because they could not get the five 120-foot spherical tanks necessary for each ship.

The South Carolina firm which had subcontracted to build the tanks said they could not deliver because of technical difficulties. When the tank production subcontract was offered for sale worldwide nobody would touch it. Finally, with the very existence of the shipyard hinging on these ship contracts, General Dynamics took over the facility and is attempting to build the tanks themselves.

For peace of mind, let us assume that they will resist the temptation to cut corners.

But will the shipbuilders of all the other nations, nine of them at present—Japan, Belgium, Norway, Spain, Ireland, Britain, Sweden, Italy, and France—resist the temptation to cut corners?

What of the LNG carriers that sail under the flags of convenience?

The Jones Act remains the single piece of legislation to date that has kept flag-of-convenience ships from trading along the U.S. coast. It requires any vessel trading interstate to fly the U.S. flag. If the energy industry's perpetual lobbying to repeal this law ever succeeds, you can be sure that overnight what little is left of the dwindling U.S. coastwise Merchant Marine will be flying the Liberian flag.

Alarmingly in line with the precedent-setting nature of the LNGCs, the Jones Act was "temporarily" suspended in February 1977 for three LNG ships. Two of the three are registered under flags of convenience, one Liberian and one Panamanian.

Industry and government sources assure us that such steps are necessary to assure adequate supplies of gas during the energy shortage. Owners of empty LNG tanks have spent millions of dollars via the media trying to convince the public that unless their tanks are filled there will be disastrous shortages. No one denies that there have been shortages, but a good many people know that some of the reasons have nothing to do with the empty liquefied natural gas tanks.

Columnist Jack Anderson says that the Federal Power Commission has been "strangely reluctant" to crack down on producers for withholding natural gas, available in this country, from the market. "But then," he says, "the illicit romance between the FPC and the oil and gas industry is an old story."

The bitter cold of 1976–77, the worst winter in 177 years, will be long remembered. Schools and factories closed; the elderly froze to death in their beds. Of all the fuels in supposedly short supply, gas got the biggest press.

Very little notice was taken of the fact that during this time the United States continued to export liquefied natural gas to Japan.

Chapter VII

THE ALBATROSS

The New York *Times*, with a Halifax, Nova Scotia, dateline, gave this front-page coverage to the sinking of the White Star steamship *Atlantic:*

"The news of the awful disaster sent a thrill of horror through the city. . . . The *Atlantic* had attempted to make Halifax Harbor on her way from Liverpool to New York, in consequence of a shortness of coal. A heavy gale prevailed at that time, so as she neared the coast, in hopes of sighting the light, the vessel was relentlessly carried by the wind and by the strong current that always prevails in that locality, right to shore, causing her to become a total wreck. . . . The steamer bumped on the rocks two or three times as the heavy waves lifted her, showing that her doom was sealed. . . . Some 250 men succeeded in getting safe to land, but, shocking to relate, none of the women and children escaped alive, all going down in the raging sea. Fully 700 men, women, and children found a watery grave. . . . The captain evidently had been running well north to shorten up his longitude. This practice had long been discountenanced by the principal lines, and in some instances positively forbidden."

The year was 1873. In the hundred-plus years since that account was written, not only has journalistic style changed but operating procedures have become more sophisticated, weather information is more accurate, communication has improved greatly, and ships are better able to withstand heavy weather.

And still the ships go down. On an average, one vessel is lost every day. U. S. Coast Guard spokesmen attribute more than eighty-five percent of all maritime accidents to human error. This is the albatross that hangs about the industry's neck.

They called her a floating art gallery, this luxury liner that was sailing, New York bound, through calm seas on July 26, 1956. The night was foggy, but she was splendidly equipped with the latest navigational aids, including two radars, one a Raytheon Pathfinder with ranges from one to twenty miles. The foghorn since midafternoon had been set to drone its warning with monotonous regularity every hundred seconds.

Docking was scheduled for early the next morning. No formal parties were scheduled, but first-class passengers were dancing in the Belvedere Room shortly after eleven and the orchestra was starting to play "A Rivederci Roma," the same tune they had been playing all evening.

This time they didn't finish.

One of the women in the ballroom said later, "I happened to look out the window and there it was, another ship right on top of us. I could see the lights and everything." Someone else said, "It looked like a city, all lighted up."

Colonel Walter G. Carlin and his wife had decided to forgo the dancing, as well as an invitation to have a drink with friends, in order to retire early. He was brushing his teeth when the crash came, knocking him to the floor. Dazed, he picked himself up. In his stateroom where, just moments before, he had left his wife reading in bed, he saw a gaping hole open to the night. The bed and his wife had vanished.

Fifty passengers were lost on that calm summer night

when the *Andrea Doria* was rammed by the *Stockholm*. The bow of the Swedish ship crushed cabins, penetrated almost a third of the width of the Italian liner's beam, and made a forty-foot gash on the vessel's starboard side.

For hours, like a broken figurehead, the body of Mrs. Carlin was impaled on the badly damaged bow of the *Stockholm*. When a crew member of that ship tried to retrieve it, he reached for the outstretched arm. It came off in his hand. Many hours later, by then full daylight, the body fell with the now detached bow wreckage and another attempt was made to retrieve it. But sharks could be seen moving in those waters. None of the fifty bodies was ever recovered.

The crash had occurred at eleven twenty-two. Within half an hour the *Andrea Doria* had a list of twenty-five degrees. The fog dispersed now, letting the moon and stars shine through when the passengers, some in evening dress, some covered by whatever they could grab, slid down the tilting, oil-slick decks into lifeboats from ships that had answered the S O S. The *Stockholm* was one of the rescue vessels that took survivors to New York.

Eleven hours after impact the *Andrea Doria*, with eleven watertight compartments which, according to her designers, guaranteed she could stay afloat, sank in 225 feet of water off Nantucket. She has never been raised.

All the facts of this tragedy will never be clear. Considerable question was raised as to whether the *Doria* was properly ballasted, indeed, as to whether she had been properly designed for stability in the first place.

The Italian vessel's spokesmen said the ships would have made a safe starboard-to-starboard passing if the *Stockholm* had not turned to her right and crashed into the *Andrea Doria*. The *Stockholm*'s claim was that they would have passed safely port to port if the *Doria* had not made a left turn, crossing the *Stockholm*'s bow. The vessels were both traveling on Track C—"Track Charlie"—one of several tracks designated by the Atlantic Trade Agreement, but not compulsory. If each vessel had been where she was supposed to be they would have been twenty miles apart.

Both the liners were equipped with radar which was manned at the time, although interpretation of the blips entailed arguments during subsequent hearings.

The masters of both ships were competent, with excellent records. Captain Calamai, of the *Andrea Doria*, had made the Atlantic crossing as master nearly one hundred and fifty times. A broken man after the disaster, he was quoted as saying, "When I was a boy, and all my life, I loved the sea. Now I hate it."

He never sailed again.

The two steamship lines at first sued each other and finally settled out of court. It will never be settled officially who was the chief culprit, but the consensus was that it was human rather than mechanical failure.

Since the two ships were of foreign registry and the accident occurred outside U.S. waters, the Coast Guard had no jurisdiction. But such a cry of public outrage was raised in this country that a committee was appointed to study the circumstances and submit recommendations for future regulatory action. It was obvious that since radar's inception there had been a missing link between the sighting of an oncoming ship and a clear understanding of her intentions.

The findings of the committee, calling for increased use of bridge-to-bridge radio communications, were submitted six months later. It was not until sixteen years later that regulations were signed which would put these recommendations into effect.

Meanwhile, in an attempt to provide this missing link, the Coast Guard set up the Harbor Advisory Radar (HAR) system for the San Francisco Bay Area as an experiment to investigate the workability of such systems, particularly in busy U.S. harbors.

When vessels report by radio, HAR, maintaining a round-the-clock watch, responds, giving information concerning other vessels. Participation is strictly voluntary; no vessel is required to report its position or to monitor the navigation radio channel, VHF channel 18A.

As a prototype of what the system could be, the San Francisco HAR seemed to be very successful. And then, early in the foggy morning hours of January 18, 1971, the Coast Guard observer, unable to make contact, watched the radar screen as two blips headed for each other near the Golden Gate Bridge in San Francisco Bay.

The blips represented two Standard Oil tankers, *Oregon Standard* and *Arizona Standard*. The *Oregon*, loaded with 100,000 barrels of Bunker C, a heavy fuel oil, was assisted by two tugs on her way out to sea and was sounding fog signals at two-minute intervals. The *Arizona* was heading for the Standard Oil dock in the bay.

Instead of keeping to the starboard side of the channel as required by the *Rules*, the master at the conn of the *Oregon Standard* elected to steer in the middle under the center span of the Golden Gate Bridge. When he saw the blip of the *Arizona Standard* on radar and tried to make radio contact, through error he set the switch on the wrong channel. Failing to make voice contact, he put his efforts into trying to avoid collision.

Now, through the fog, the lights of the oncoming *Arizona* were sighted visually, the white and green navigation lights on the starboard side, only two hundred and fifty yards away. He ordered full astern and sounded the general alarm.

Meanwhile, the master of the *Arizona* had been trying without success to contact the *Oregon* by radio. Too late, he saw off his starboard bow the red portside lights of the *Oregon*. He ordered hard left rudder and the engine stopped.

Almost directly under the Golden Gate Bridge and almost softly, the two vessels collided. The shock of impact was so slight that not one of the men on either vessel was injured or even lost his footing.

But the result was an ecological catastrophe. The *Arizona*'s bow had penetrated the port side of the *Oregon*, rupturing three of her cargo tanks. Locked together for seven hours, the vessels drifted toward Angel Island; 840,000 gallons of oil spilled and the mess was carried by the tide into San Francisco Bay and the adjacent coastal areas.

This was Bunker C, a fuel oil so heavy that it must be kept heated by submerged steam coils. It can never be allowed to fall below 130 degrees, since at that temperature it congeals.

When the hot oil poured out of the ruptured tanks and struck the cold waters of the bay it became a solidifying, tarry mass. It spread, clung to beaches, boats, and birds.

Within a short time Standard Oil was doing everything possible to clean up the mess and keep it from spreading. Vacuum hoses sucked up the clotted oil from the surface. Helicopters dropped tons of straw and cranes on barges picked it up, as did men in dozens of small boats. Volunteers eagerly accepted the challenge and waded into the water to collect armloads of the gooey stuff, stacking it in great piles to be trucked away.

But the poor waterfowl sank helpless, unable to fly with their oil-gummed wings, cold because their protective down was sodden. Volunteers gathered hundreds of them and hurried them to cleaning stations. Standard Oil provided thousands of gallons of the kind of oil that is used to clean the delicate skin of babies. Bathed repeatedly in this, then dried in a desiccant mixture of flour and corn meal, the birds were force-fed with nutrients, antibiotics, and vitamins.

But 96.5 per cent of them died.

The Coast Guard Marine Board of Investigation summed up their evaluation of the accident by saying, "The casualty was caused by faulty navigation of the SS *Arizona Standard* and the SS *Oregon Standard*. Both vessels proceeded at an immoderate speed in dense fog and failed to keep to the starboard side of the channel prior to the collision." Other factors noted were failure of the *Oregon* to make timely radio contact, and the loss of radar contact by the *Arizona*.

As in the cases of the *Atlantic* and the *Andrea Doria*, human error was primarily responsible.

The competence of the individual mariner received no legislative attention in this country until 1838. Since then, significantly, laws have followed disasters.

▶ The Act of 1871, embodying today's safety code, was passed after the *Sultana* sank on the Mississippi River in 1865 with a loss of 1,500 lives. That vessel, certificated to carry a maximum of 365, was loaded with 2,300 persons, most of them Union soldiers returning home after the end of the Civil War.

▶ The International Ice Patrol was the result of the loss of the *Titanic* in 1912—the ship that "God Himself could not sink."

▶ A bill was passed placing on vessel owners the duty to ensure "that the crew is fully trained to meet all emergencies due to fire, collision and stranding" after the 1934 burning of the *Morro Castle*.

▶ The Bridge-to-Bridge Radio Telephone Act was an outcome of the sinking of the *Andrea Doria*.

Human error can never be legislated out of existence. Man will never be infallible. All the gadgetry and refinements available now and in the future will never make him so.

During the fiscal year 1975 the United States Coast Guard investigated 3,305 marine casualties. Three hundred and twenty-five vessels, almost one a day, were declared total losses.

Are there factors aboard ship that contribute to the commission of error? Are there stresses not to be found in other occupations? What is life really like out there?

Several shore-based friends have said to me, "I've always wondered what your life was like at sea—and now I know. I've just read that book by what's-his-name—anyhow, a great book, *Supership*."

"Right. Noël Mostert. But the ship he sailed on was British, and the British Merchant Navy is not really the same as the U. S. Merchant Marine."

No. It is not really the same.

Mostert wrote: "The principal diversion of the day was sundowner time when the working day was over for most, enabling the juniors to gather in the wardroom and the seniors to start their pour-out in one of the ship's several executive suites."

And: "They sat on the black leather chairs and sofas wearing their beautifully laundered dinner kit . . . and in their hands were tumblers of eight-year-old Glenfiddich whisky which the barman Laurence silently refilled at the briefest glance."

He described a gourmet dinner menu which ended with "peach melba and the savory such as bloater paste on toast. 'Who is going to be Mum?' someone invariably asked, and whoever was nearest to the coffee began pouring." He said that "A normal breakfast menu might be stewed apples, cornflakes, oatmeal, smoked cod in milk, sausage mince cakes, fried potatoes, cheese or plain omelettes, eggs to order, rolls and toast, tea or coffee."

On a typical U.S.-flag ship you come down to the cafeteria each morning and there on the chalkboard is the breakfast menu: eggs (you can hope to get them cooked the way you want); chipped beef on toast, appetizingly written on the board as S O S (shit on a shingle). Canned juices are available, as are other beverages, indifferently prepared. If you should want an extra piece of toast you may get the toaster shoved at you across the counter: "Make your own goddamn toast."

This is not to say that the company does not buy good food and plenty of it. But the cook may be somebody who couldn't make it on deck or in the boiler room. To be a ship's cook requires no culinary skills, and often what he does to good food shouldn't happen to a dog. Many of the men choose to cook in their own rooms. Or to snack from the refrigerator on ships where that is permitted.

We have no "pour-out" time in the "executive suites." Liquor is illegal aboard U.S. tankers and there are no executive suites. Mostert wrote of crisp linens and silver tea services and uniforms being worn "of course." Crisp linen and silver tea services are unheard of and uniforms are not issued or required. It is not uncommon for an engineer to come straight from the engine room, covered with grease, and seat himself for a meal. I have known seamen to put on and wear clothes they found in the bag of rags that came aboard for engine wiping.

Aboard a British tanker "The swimming pool and cinema are no longer regarded as luxuries, but as indispensable items of tanker life." Their staterooms are equipped with double beds for the times when wives choose to sail with their husbands.

I know of no U.S.-flag tanker that has a swimming pool. Few have movies. Officers' wives are not allowed to travel on board our tankers, with the possible exception of the captain's wife.

We have no luxuries, no amenities, no rituals aboard a U.S. tank ship. Not only is gracious living by-passed, but very often so is common decency.

So what is the lure?

Money. U.S. seamen are the highest paid in the world, receiving almost four times as much as the British. Our ordinary seaman generally works between six and eight months a year, during which time he earns from fifteen to twenty thousand dollars. A chief mate can make as much as fifty thousand dollars for six months' work and he can double that if he chooses to work full time.

Whether the British seaman would be willing to by-pass his life style for the higher pay is not for me to say. But I feel safe in saying that the American would not be willing to trade.

What makes him tick, this American who chooses to go to sea, embracing a life that statistics show to be among the most hazardous? Is he the stereotyped adventurer in search of excitement in foreign lands, a carefree salty dog with a girl in every port?

Until 1971 the Merchant Marine industry did not have the most basic information about its human resources. What were the backgrounds of the officers who sailed in the U.S. merchant fleet? What were their initial reasons for going to sea? Why did they remain?

The Personnel Research Division of the National Maritime Research Center did a study of several hundred American Merchant Marine officers and came up with a profile that looked like this:

Quite a number of them are now middle-aged; they are married, with several children. Most of them were born in coastal states, with a background of average economic status, a record of average grades in school. About a third attended college and half of those secured degrees. Their hobbies and interests tend to center around solitary pursuits.

The romance of the sea and the desire to travel were major reasons for going to sea, but these reasons apparently have little or nothing to do with staying there. Salaries and job benefits of various kinds take precedence over all other reasons for remaining. The single outside activity cited more often than any other has to do with looking after their financial investments. Few find the seafaring job itself to be challenging or even interesting.

Evidence of low morale is shown by these statements

from the study: "Licensed personnel report that conditions under which they must work, the type of people with whom they must work, and the status and insecurity of their jobs, have deteriorated to the point that the job is no longer pleasurable or even bearable. The over-riding impressions that emanate from these responses is of desperately unhappy and dissatisfied men. . . .

"According to their responses, if the men surveyed had their lives to live over, three-quarters of them would choose a non-marine related occupation. Only 1.7% would choose the same career they have now. . . .

"A fairly common theme which seemed to emerge from the Merchant Marine Attitude Survey was the relatively low esteem in which respondents felt their maritime positions are held, a theme expressed as a generalized plea to 'somehow' improve the status of the Merchant Marine."

In a rather sad attempt to build a more acceptable image shoreside, some of those with lower ratings paint "Captain" on their mailboxes, stick "Captain" stickers on their duffel bags once they're out of sight of their shipmates. Not long ago in an airport I wondered why a third mate I knew well was so ill at ease, and then I saw his duffel with the "Captain" sticker which, apparently, he hoped I would not see. Mail arrives aboard ship addressed to a captain who may be an oiler. It's from a girl friend, maybe even a wife who doesn't know he is not a captain.

It works the other way too. At sea, to hear some of them tell it, they're big shots when ashore. One told us he was an ex-national champion motorcycle rider. . . . Hm-m. It turned out he wasn't, and was terrified of even riding a motorcycle. Another said that, in 'Nam, he was a Navy flyer, flew an F-4D Phantom until he had a flameout and broke his back. I used to sit and listen and nod my head even after I found out the Navy never flew the F-4D Phantom.

We have many reminders that mariners are held in low esteem. In port cities I have seen signs saying, "Painters

Wanted—Seamen Need Not Apply." Legal phraseology provides other put-downs, traceable to medieval sea codes, but still applicable today and on the books. The seaman is considered to be a ward of the court together with infants, incompetents, and the insane. In maritime law, he is described as "poor and friendless" and apt to acquire "habits of gross indulgence, carelessness, and improvidence." Also, he is "endowed with invincible ignorance."

Those who are not officers are called "unlicensed," which carries its own derogatory connotation. "Oiler" and "wiper" are actual job designations, and that's what you write in the space marked "Occupation" when, say, you apply for a bank loan or car insurance or fill in the space for father's occupation when your child is born.

A slang term for the engineers is the "black gang," a holdover from the days when coal was burned in the engine room. The ship is referred to as the "zoo," and terms like "deck apes" and "jungle bunnies" and "animals" are in common usage and occasion no surprise.

A group of psychologists from the University of Maryland went on a five-day freighter cruise to get the feel of "having been there." A worthy objective—those firsthand observations and face-to-face queries no doubt contributed valuable information to the study they were making for the union-sponsored Harry Lundeberg School for Seamanship. But a million questionnaires could be computerized and their spewed-out answers would never really give the feel of having been there.

Being there means that you sometimes work for thirty-six hours straight, literally falling asleep standing up. You work alongside men who seem not to bother to bathe from the beginning of the voyage to the end of it. You risk your life cleaning gas-pocketed tanks. You freeze in the icy gales of the North Atlantic and collapse with heat prostration in the Persian Gulf. You carry unseen cargo from nowhere to nowhere—a submersible pipeline five or ten miles offshore to a man-made cement island.

Turnaround time is short; often the few hours you might spend ashore are denied because there is no access to shore. No wonder the mystique and romance and adventure of faraway places don't figure when men check off their reasons for staying with the sea.

A major reason for disenchantment is the disruption of normal home and family life.

Take my friend Hank, a first mate. He worked extra time. For fourteen years he never missed sending his pay check home. That house in Hawaii had a swimming pool, everything. Nothing was too good for his wife and kids.

It wasn't enough.

His wife spent the money, but not on house payments. The bank foreclosed. He told me, "My wife's shacked up with some guy. My kids don't know me. I hear my son's doing drugs and my little girl is a tramp." Hank spends almost all his time at sea now. He drinks a lot. As first mate, he's supposed to run the ship, but I've seen times when we would have had collisions if others hadn't been there to cover for him.

A British study came up with the conclusion that most collisions were attributable to negligence and appalling seamanship.

Shell Oil's study said it more simply: "People make silly mistakes."

The silly mistakes that people make can be triggered by every emotion, every variety of psychological and physiological stress.

Boredom. The most stultifying, mindless job in the world surely is rust chipping. You go chip, bang, chip with your hammer day in, day out, knowing that the rust breathes over your shoulder, waiting to attack the fresh skin of the metal you've just exposed. Keeping lookout is boring too, most of the time. A sort of highway hypnosis sets in. When an emergency erupts you can't always change gears fast enough to react effectively.

Fear. You're a pilot, aboard a liquid gas carrier—let's say it's the Swedish *Claude,* carrying 900 tons of liquefied butane, and you collide with a freighter off Southampton, England's leading passenger port. Seconds after impact, you find yourself alone on the bridge, the ship's crew having leaped overboard. You follow, not knowing the cargo, but figuring they knew something you didn't know. The *Claude,* still running without a soul aboard, goes aground. She is towed to a refinery where a Portuguese ship agrees to take off the liquefied gas. A cargo hose ruptures, releasing a gas cloud, and now it's the Portuguese tanker that panics, puts her engines on full speed ahead, causing other hoses and pipelines to tear apart. Volunteers risk frostbite as they try to close the valves.

Spite. You have a tyrant for a first mate. Hoping to get him fired, and knowing he will be held responsible, you and the other unlicensed seamen "deep six" the vitally necessary Butterworth tank-cleaning machines. So upon arrival at Ra's at Tannurah, Saudi Arabia, the ship is unprepared for loading and has to wait while new machines are air-freighted 12,000 miles. The first mate is furious, and now he tries to make up for lost time by driving hard when they're under way again, endangering the safety of the ship and the crew.

Depression. December 18, 1971. It's a cold predawn morning on the tug *Maryland,* and the winds are high as you enter Albemarle Sound, North Carolina. You've just been demoted from master to first mate for no fault of yours, and it eats at you; you know the man who has your job isn't half as qualified as you are. Something comes up that should be reported to him and you say, what the hell, and don't report. Six of the seven crew members die, you among them.

Thirst. A fire breaks out in the engine room. Somebody grabs the fire extinguisher and it goes pffft, nothing, because the cold CO_2 has been used to chill drinks. It's a shocking thing to think that the fire extinguisher is empty for a reason like that. But there you were in the engine room with the temperature at 130, retubing the sooty insides of a boiler, all carbon and sweat . . . and there sits a canned drink, warm, and you shoot it with the chilling spray because right then your parched throat is all you can care about.

Tension. The alarms on the bridge monitor console go off, all the bells and the beepers and the buzzers, and you're going nuts. You grab the hammer that lies there—some smartass has marked it "engine alarm reset tool"—and you smash at the console to shut everything up. They malfunction most of the time anyhow, and the sophisticated console with all the latest gadgets has been slammed so many times it looks as if it's been antiqued on purpose with a ball-peen hammer.

Negligence. You are preparing to discharge cargo and you tell the "ordinary" to go warm up the cargo pump engine. It is huge, a twelve-cylinder diesel, housed in a watertight, airtight enclosure. You *assume* the air intake doors are hooked open. They are not. The engine starts, almost immediately exhausts the air in the tiny space, sucks the access door shut, creates a vacuum, and the ordinary is trapped inside. The access door and the intake doors are impossible to budge, sealed. You think of brain damage, death. As chief mate you are responsible. Not a moment too soon you are able to open a drain valve and the trapped seaman, on the verge of collapse, crawls out.

It so happened that he was the captain's son. I remember it well because I was the chief mate.

Examples could go on and on, enumerating every defect of character, every silly little mistake, every weakness to which humankind is heir. Commander Benjamin E. Joyce, U. S. Coast Guard Chief of Manning and Personnel, pushes the percentage of accidents caused by human error to a possible ninety-five per cent.

Officers list drinking as a major problem among their

crew, but alcoholism is by no means confined to the crew-men.

I have seen a captain leave port, drunk out of his mind, and wake up at sea a day and a half later, not knowing where he was.

I have seen a chief mate hauled like a sack of sand out of the way during docking. Entering and leaving port are the two most crucial times during the voyage and this is when drunkenness is most common. This particular chief mate was rewarded with a captaincy when he reported in to the company offices, which happened to be thousands of miles away. Vessel management teams seldom know their captains or any of the rest of the crew, for usually they have been ar-bitrarily assigned by a union hiring hall.

I was second mate on a new tanker that sailed worldwide, and at every port the captain would buy liquor out of bond —the finest of whiskey, champagne, and beer. Charged to the company and listed by the steward as detergents . . . brooms . . . meat . . . it was sold by the captain to officers and crew members. He installed a beer machine in each mess hall. Whether we wanted it or not, beer by the case was brought to our cabins and the cost was deducted from our wages.

There are men at every level in the Merchant Marine who are capable and conscientious, and they are just as con-cerned about these problems as I am. If I seem to have stressed the incompetents, it is only because they have a disproportionate importance now. In April of 1976 the Todd Shipyards Research and Technical Division said in a report sponsored by the U. S. Maritime Administration: "With the development of large crude carriers, and larger and faster express service ships and LNG carriers, the consequences of collision become more serious."

We are told that men who are responsible for the LNG carriers will be specially selected, specially trained. I hope that is true. They may have a band of angels waiting out there. I only know what I see.

A captain of my acquaintance has acquired, with middle age and many years of experience, an easygoing manner, but he is skilled, one of the best. He sails on one of this country's largest chemical carriers, a tanker. In addition to the conventional tanks below decks which hold a variety of so-called "exotic" chemicals, the company, to increase carrying capacity, has installed above deck a self-supporting LNG tank, welded into place. This does not make the vessel an LNG carrier, as such. The tank is small, comparatively, and called "portable," but this does not mean you can pick it up and carry it. It holds 84,000 gallons, or the capacity of an average-size swimming pool.

This tank, my captain friend told me, has leaked from the day it was first filled. When the loading arms are attached to the manifold there are leaks of LNG at the bolted flange connection. To stop the leaks, wet rags are thrown over the leaking area, then water is squirted, freezing on contact and forming a makeshift ice seal around the connection.

We were sitting having coffee while he was telling me about this. A statistic came to mind, one included in a congressional hearing: in a five-year period liquefied petroleum products were involved in only about ten per cent of all pipeline accidents but caused nearly seventy per cent of all reported deaths.

I asked him if anyone was concerned enough to have this situation corrected.

He gave a shrug. "Sure. The whole ship's force—at least the ones who know something about the dangers of the stuff. It's much too sophisticated for anyone aboard ship to fool with, so what can any of us do about it? Refuse to sail? Yeah. With ten men waiting in line out there to take every job."

"And you've reported this to the company—"

"Oh, of course. Listen, when the ship first came out there

was no drip pan under the tank. So the first time the relief valve opened—which it has to do when pressure gets too high—the liquid dripped onto the deck. Immediately we had a four-foot opening—not just a fracture. Lucky for us, the deck split over a cofferdam."

(The cofferdam is a void space in which cargo is never carried.)

"But if that supercold stuff had ever fractured the tank of one of those caustic poisons down below—who knows what would have happened? I guess nobody ever will know until a thing like that happens. It'd be a helluva way to find out. Anyhow, what they did was to withdraw the ship from service to repair the deck. This time they put a drip pan under the tank."

"Oh. Well, good."

"Sure. But what happened, the high-level alarm wouldn't stay silent anywhere but in port. When we were at sea the liquid sloshing around in the almost full tank would sound the alarm every few seconds. Everybody was going bananas. So we silenced the alarm."

That figured.

"So the next time the LNG tank was being filled the high-level alarm didn't go off, *couldn't* go off, being disconnected. And the pressure built up until the safety relief valves lifted and let some of the LNG out. Not much, some."

"Into the new drip pan."

"Yes. Into the new drip pan. *But* somebody forgot to put the drain plugs in place. Out comes the LNG just like before, down onto the deck, opens up another hole. Down into the cofferdam it goes. Just like before. So back to the shipyard we go for a couple more new deck plates."

I asked what would happen if the pressure inside the tank climbed so high the relief valve could not release it quickly enough.

"Oh, the rupture disc gives way and lets the entire contents of the tank spill out."

"And the drip pan isn't large enough to contain the whole 84,000 gallons—"

"Hell, no."

"How much would it hold?"

He had no idea. Very little.

"So if all this liquefied natural gas spilled out onto the deck—?"

He said, well, he didn't think it would happen.

I asked him if he had ever had any special training concerning the chemicals or the LNG before taking over this ship.

"No," he said. "None. . . ."

Chapter VIII

THE TOUGHEST PROVING GROUNDS ON EARTH

> The sea, like a great sultan, supports thousands of ships, his lawful wives. These he caresses and chastises as the case may be.
>
> Felix Riesenberg, *Vignettes of the Sea.*

December 8–13, 1969, Seattle, Washington. The *Badger State*, a 441-foot U.S.-flag cargo ship, loaded 500-, 750-, and 2,000-pound aerial bombs round the clock. When cargo was loaded, vessel repairs had been made, and she had received her U. S. Coast Guard annual inspection, she was "ready for sea."

December 14. In the early evening the *Badger State* departed with captain and thirty-nine crew members westbound for Danang, Republic of Vietnam.

December 15. Heavy weather was encountered immediately upon leaving sheltered waters and entering the Pacific Ocean. The Naval Control of Shipping Organization, which controlled optimum safe weather routing for the voyage, sent a diversion order directing the *Badger State* to steer southwest to avoid heavier weather.

December 16. The ship complied with the diversion order. Rolling to forty degrees in huge seas continued. Some shifting of the bombs occurred, but the crew made necessary adjustments to stowage.

December 17. Rolling increased to forty-five degrees. A second diversion order was received directing the *Badger State* to steer due south. Crew members worked through the night to resecure the shifting 500-pound bombs in #3 hold. At the same time engineers worked to fix a broken steering mechanism.

December 18. Bombs began breaking loose from their wooden pallets in four other holds, resulting in steel-on-steel contact with each other and the sides of the ship. Strenuous crew efforts continued until by midmorning refastening was completed. A third diversion order directed the ship's course to the west. The winter storm continued to rage.

December 19. A leak in the *Badger State*'s hull was discovered, a wasted area below the propeller shaft. Cement was poured in to patch the area, and speed was reduced to prevent the repair from vibrating loose while the cement hardened.

December 20. Continued rough weather with no letup necessitated constant inspection and repair to resecure breakaway bombs.

December 21. Wooden dunnage boards, used to keep the bombs in place, were being splintered and broken faster than the fatigued crew could replace them. The storm continued.

December 22. A fourth diversion order directed the vessel's course west-southwest. Twenty-foot seas and thirty-five-degree rolls forced the *Badger State* to heave to in order to continue wrestling loose bombs into place. By now, dunnage boards to make repairs were getting scarce. Bombs were adrift in every hold. The crew was exhausted.

December 23. The master radioed, "Request urgently that you get me south to good weather as soon as possible." A fifth diversion order for a southwesterly course was soon received. Frantic efforts to resecure rolling bombs in every cargo compartment were frustrated as a single bomb would loosen up several more on impact.

December 24. A sixth diversion order came to make for Pearl Harbor. The ceaseless cycle of stormy weather, violent rolling, bombs breaking loose, and efforts to resecure continued, with the crew desperately, sleeplessly, hanging on.

Christmas Day. At 2:00 A.M. the *Badger State* encountered a much more severe storm, unpredicted, with winds of hurricane force and mammoth confused seas. Rolling to fifty degrees, bombs were loose everywhere. The hurricane raged for the rest of the day and night. By now the master had been on the bridge almost continuously since the twenty-third. The crew also had worked nearly round the clock without stopping.

December 26. A seventh diversion order came on the twelfth day of the *Badger State*'s fight for her life. She was directed to Midway Island, the closest port. Now a second unpredicted storm hit. She was hove to when she encountered a mountainous sea, powerless to make any headway against the storm. Efforts were made to launch the port lifeboat, but the seas carried it away. Likewise, the vessel's only two inflatable life rafts were lost.

Half of the hatch cover was winched open and 2,000-pound bombs were seen sparking in the darkness of the holds as they slammed into each other against the ship's side. The crew began throwing mattresses, linens, towels, reefer stores, and eventually even frozen meat and other food supplies into the hold, hoping that these materials might help cushion the impact of the bombs. One crew member hysterically tore off his clothes and threw them in.

Everybody knew that if just one of those bombs went off it could be all over.

At last the captain of the *Badger State* sent his Mayday call. Forty miles away, the distress message activated the autoalarm on the *Khian Star*, a Greek merchant ship, who radioed back that she was coming to assist. As the Greek ship came into sight, a violent explosion occurred in the #5 hold, blowing off the hatch cover and making a large jagged hole in the ship's starboard side.

Immediately the master sounded the abandon ship signal and the general alarm. He and four volunteers elected to stay aboard, hoping they could get the *Badger State* to a safe port.

Thirty-five men climbed into the lifeboat and struggled to release the falls that were going slack as the boat rose on the crest of the mountainous waves and then snapped tight each time she fell back into the troughs. As the boat became water-borne, the swells were so high that they would carry it back up to the boat deck before the falls could be released. Despite the efforts of the men working the hand-propelling mechanism, the boat drifted aft. One wave slammed them against the torn area where the bomb had made a hole and the jagged metal gashed the head of one of the men.

When the boat was abeam of the hole in the #5 hold, a 2,000-pound bomb rolled out, as if from a bomb bay, landing squarely in the lifeboat, killing and injuring several of the crew. The boat overturned; some of the men were able to cling to it as it drifted away from the *Badger State*. At this point the captain and the four crew members remaining on board with him went over the side into the rough cold waters.

Now came the albatross on their "obscene wings" which stretched to as great a width as twelve feet. Known to be meat eating, and held traditionally in superstitious fear by seamen, they attacked the men struggling in the water. The

Khian Star, herself rolling thirty to fifty degrees and with large swells pouring over her decks, moved closer, hauling one man aboard by a lifeline which had knotted around his ankles. Another was hoisted aboard tangled in a cargo net. Eventually fourteen survivors were picked up by the *Khian Star.*

Search vessels stood by and watched the intermittent white, red, and orange flashes aboard the *Badger State* for the next nine days and nights.

January 5. The *Badger State* sank.

LNG ships carry their own unique loads of improperly secured "bombs," those thousands of tons of liquefied natural gas sloshing back and forth as the ships roll in heavy seas. The *Polar Alaska* and the *Arctic Tokyo,* both LNG vessels, and both designed by Gaz Transport, Paris, had tank-wall failures. Audy Gilles, managing director of that company, attributed the failure to sloshing and said, "The 'slosh effect' results in very violent impact against tank walls."

The chief engineer of one of the largest builders of LNG ships told me, "We've got a real problem and we don't know exactly what to do. These are dynamic forces. We just can't afford, though, to wait for technology to catch up."

The standard size LNGCs have another problem that is more serious than for a vessel the size of the 441-foot *Badger State.* Less than half the length of the LNGC, the *Badger State* could behave more like a cork, her hull almost always near the surface as she climbed the mammoth swells, pitching and rolling to fifty degrees. The LNGC's shell, almost a thousand feet long, cannot bob, but instead must span the swells, leaving parts of her loaded hull practically unsupported between wave crests.

When the ends of a vessel rest on two successive wave crests with no support to the mid-section the hull strain known as "sagging" takes place. In the next moment she may be subjected to another massive strain, known as "hogging," when the midsection is supported by one wave holding her aloft as if she were on a pedestal with her ends suspended out over the troughs. Sometimes as a result of such extraordinary wave situations a vessel will snap, her hull tearing apart at the rate of 5,000 feet per second.

"The *Fitz* is gone!" The captain of the *Arthur M. Anderson*, who had been watching the *Edmund Fitzgerald* on his radar screen, delivered that terse message by radiotelephone to the Coast Guard on the stormy night of November 10, 1975. Her demise was so sudden that no distress call was heard from her before she plunged 535 feet to the bottom of Lake Superior.

It was snowing. The "witch gales" of November were blowing eighty miles per hour and thirty-foot waves prevailed at the time of the accident. The *Fitz*, loaded with 26,216 tons of taconite pellets (iron ore), had lost two ventilator covers, was taking on water, and had developed a list. But voice communications prior to the sinking showed no undue concern:

The *Anderson:* "We haven't got far to go now—we'll have it made."

The *Fitzgerald:* "Yes, we will." And minutes later, "We're holding our own."

The skipper of the *Anderson* testified at a Coast Guard Board of Inquiry investigation that he believed the captain hung up the phone after those words and "immediately the *Fitz* dove right under. I don't think those fellows even had time to get out of the wheelhouse."

Great Lakes vessels have long been considered floating coffins because they don't have watertight bulkheads. If the water reaches the cargo hold the ship must go down.

Not one of the twenty-nine crew members has ever been found. The ship's bow and stern sections have been photographed extensively by a Navy robot vehicle, but no photograph shows a body.

Lake Superior, so legend has it, never gives up her dead.

Underwater photographs show that the ore carrier snapped at the midsection, probably having been caught between two giant waves which broke her back. Experts speculate that the stern half, with the propeller still driving her hard, smashed into the mud bottom at a speed of about thirty knots. Coast Guard photographs taken by an unmanned diving bell show that section lying upside down. The bow section is sitting upright nearby, still intact except for her jagged ends where the hull was sheared and torn in two so suddenly.

Gordon Lightfoot wrote a song, "The Wreck of the *Edmund Fitzgerald*," which very quickly reached the golden-record mark, with sales of a million copies.

Three years before the wreck of the *Fitzgerald*, I had my attention called to a similar situation when my professor of tanker operations held up a newspaper photograph to the class. It showed a new tanker, the *Martha Ingram*, upended, her hull split across the middle like a drawbridge, her bow and stern sunk below the water, resting on the mud bottom alongside her dock in Port Jefferson, New York.

"Obviously," he said, "somebody goofed." It seemed to illustrate perfectly a point he had made to us over and over about improper loading. But this was not the case. Subsequently we learned that all of her cargo had just been discharged.

The 33,000-ton *Martha Ingram* was only a year old, and built to the highest American Bureau of Shipping standards. But there, alongside the dock with the sun shining and a light breeze blowing, in a fraction of a second her hull snapped in two. Dazed but uninjured, her crew made it to safety.

Three years later, almost to the day, the *Martha* entered

that harbor again on her first voyage back to Port Jefferson after lengthy salvage operations. I was on her this time as first mate and had had a chance to read the final conclusions of the investigation as well as talk to the captain who had been on her at the time of the breakup.

No one really seemed to know what had happened. A minute flaw, a deck fracture only one eighth of an inch long in the area of the king post, may have been the weak link which led to the ship's breakup.

Normally, tanker hulls are remarkably flexible. Before leaving the dock, I routinely load with eight inches bend in the hull (hog or sag) as a result of uneven cargo distribution. The amount of bend is readily visible since at the bow, stern, and amidship columns of white numbers, at one-foot intervals, are painted on the side of the hull, beginning with the lowest number at the bottom. Reading where the waterline cuts across these numbers gives me the amount of hull submersion (draft). Armed with the three draft readings, the amount of bend in the hull is quickly computed: I need only average the bow and stern waterline figures to get the mean, and the difference between this mean and the midship reading quantifies the amount of hog or sag.

For example, if the mean figure is twenty feet and the midship reading is twenty-one feet, then I know she's sagging a foot lower in the middle. Were the midship reading nineteen feet, then I'd know we had one foot of hog.

Once at sea, a vessel no longer has a fixed quantity of hogging or sagging. No longer static, she must adjust to constant flexing. After a few years at sea one becomes blasé about many things, but I seem not to be the only one who never gets used to the awesome sight of a great ship waving, humping up and down against the stationary horizon.

"Look at 'er," somebody will say, and we stand there looking down the uncluttered deck. Narrowing in the distance, she looks like a fishing pole responding to the rapid, steady nips of a fish on a line.

This working, this flexing and bending, continue for the life of the ship and can lead to metal fatigue with resultant fissure. The critical service period is estimated at eight to twelve years after build. It takes at least this long to test a hull's wearing qualities. She may perform reliably for many years and then, as overstressed areas begin to reveal themselves, start coming apart—but gradually, giving enough warning so that she can make it to the repair yard.

And then again she may, God knows, let go all at once.

March 27, 1971. Cape Hatteras reached for another victim to add to her crowded graveyard of at least 2,000 ships. This time it was the *Texaco Oklahoma,* Boston bound, with 220,000 barrels of black furnace oil. For two days, whole gale sea conditions had prevailed with thirty- to forty-foot waves making it necessary to alter course temporarily whenever a crew member had to go out on deck. The ship had been laboring in a manner described as "shuddering," and at three-thirty in the morning, with a forward pitch and a starboard roll, the twelve-year-old tanker snapped completely in two.

But the sea was by no means through with its prey.

Thirty-one men on the stern section heard the loud crack. Most of them had been asleep. Not yet comprehending the enormity of what had happened, they raced, stumbling, to lower the starboard lifeboat. There, with disbelieving eyes, they saw the detached forward section bearing down on them, tilted bow up, with what looked like a flashlight signaling from the wheelhouse. In a schizophrenic maneuver, she slammed into the readied lifeboat, completely destroying it.

Amid burning paint and sparks caused by the friction between the two parts of the hull, the stern section, which housed the engine, backed away to avoid further damage. To the crewmen on the stern, it was as if their father had turned on them, for the bow housed their captain with a dozen men.

One of those men was Third Mate Robert Paul, a friend of mine from Mexico, Missouri. Neither he nor any of the others was ever seen or heard from again.

The bow of the *Texaco Oklahoma* had carried away two lifeboats with it and the only lifeboat left for the men had been stripped for maintenance. Immediately, the men set to work making this boat ready for sea and launching. The lifeboat emergency transmitter was set up. The radioman had been on the bow section; none of the men left knew much about operating the radio or sending an S O S, but they read the instructions, rigged the antenna, and did their best. Men in pairs then hand-cranked the equipment continuously. Within a couple of hours they heard a broadcast reporting that there was a search under way for a tanker that had broken up at sea. Their signals had been received! Help would be there soon!

6:00 A.M. "A ship!" someone shouted. "I see a ship!" Sure enough, a ship was coming at them through the tossing sea. Spirits were high. And then they saw the name in big white block letters on her bow—TEXACO OKLAHOMA. Their own bow was coming back at them, headed as if to ram them again!

They put their engines full astern and got out of the way just as the bow passed close by. It was as if she were carried by some special separate current, unfelt by the part of the ship to which she had once been joined.

This time no signs of life were seen. But the lifeboats were still intact, and the men aboard cast longing eyes toward them as the broken bow drifted out of sight.

6:30 A.M. A ship! This time it was a whole one that was sighted on the horizon only eight miles away. The thirty-one men cheered and sent up flares. But the ship faded away.

They posted continuous shifts of lookouts, but no other ships were sighted during the rest of that morning.

5:00 P.M. A second ship came into view, rekindling hopes for rescue. Several flares were fired. But this too passed on out of sight. The weather had not improved and the stern section was beginning to tilt slightly forward.

Evening twilight. A third vessel appeared and this time drew closer, only five miles away. The overjoyed crew let go with what must have appeared to be a Fourth of July celebration—parachute flares, smoke signals, blinking white and red lights. They blew their whistle repeatedly. The ship signaled back by a flashing light, clearly directed at the wounded remnant of the *Oklahoma.* The crew went crazy with joy, jumping up and down, hugging one another.

The vessel came no closer. For the next two and a half hours she stood out there, not close enough to be identified, but obviously observing them. Perplexed but hopeful, the crew thought she might be awaiting the arrival of Coast Guard helicopters.

8:00 P.M. The ship turned, steamed away in the direction from which she had come.

8:30 P.M. Darkness now, and the pounding seas rose to carry away the port lifeboat which had been swung out and made ready for launching. One of the two crank handles on the emergency transmitter broke and the shaft seized, rendering the transmitter useless.

3:30 A.M., March 28. Now twenty-four hours had passed since the ship snapped in two. The trim reached thirty degrees by the head. Convinced that sinking was inevitable, the engineers began securing all machinery.

4:30 A.M. The steam-driven generator kicked out and the emergency diesel generator picked up the load. Small amounts of water began to seep in at the forward ends of the deckhouse passageway.

5:30 A.M. The pitch increased to fifty degrees. It was almost impossible to maintain footing. The captainless crew decided to abandon their half of the ship. Wearing life preservers and clutching what ring buoys had not blown overboard, the thirty-one crewmen gathered on the stern. Two rafts had been hastily improvised by lashing together empty oil drums. They threw these overboard together with the inflatable fifteen-man rubber life raft which had been secured by a line tied to the rail.

Just as the raft began to inflate, a wave swept it under the projecting lifeboat davit arm, completely collapsing the canopy and effectively blocking entrance to the inside of the raft. Fifteen men climbed down the Jacob's ladder and scrambled on top of the collapsed canopy. Before the others could join them, the line broke and the life raft began to drift away.

A cargo tank ruptured, sending out a wave of black oil which washed all the men off the rubber raft. Gagging on oil, they struggled to get back aboard the raft. Eleven made it. Four others grabbed a big board. All of the men had jumped into the sea now and were trying to catch and cling to the slippery oil-drum rafts which kept flipping over and over in the punishing waves.

6:05 A.M. The stern section upended to a ninety-degree angle and slid out of sight.

Exhausted and sick from swallowing oil and salt water, the eleven men clung to the slick top of the damaged life raft. They saw two ships and an aircraft but failed to attract attention. Finally they managed to raise the canopy and crowd inside its shelter where they found emergency supplies of food and water. More than thirty-six hours had elapsed since the *Texaco Oklahoma* split in two. The men collapsed.

5:00 P.M. A ship's whistle! The men fought their way out through the canopy door. The Liberian tanker *Sasstown* was standing by so close that the low afternoon sun cast her shadow across the life raft. Purely by accident, she had sighted the tossing raft. She had to make several passes before she was able to throw a rescue line for them to catch. One by one the men climbed up the Jacob's ladder. The next day two others were rescued afloat in their life preservers, bringing the total of survivors to thirteen out of a crew of forty-four.

The *Sasstown's* report was the first inkling the Coast Guard or anybody else had received that the *Texaco Oklahoma* had sunk.

The sea has been described as a liar, a seducer, an unmated creature, a devil, a thief, but it cannot be blamed for all the bizarre happenings surrounding the loss of the *Texaco Oklahoma*. An almost unbelievable combination of coincidences seemed to conspire against her:

The broadcast the crew members heard, reporting that a search was under way for a distressed vessel, was for one six hundred miles away and did not concern the *Texaco Oklahoma* at all. Indeed, it perhaps did not concern *any* vessel, for after an exhaustive search over an area of 11,250 square miles, assisted by a naval aircraft and three merchant ships, nothing was found and those who had engaged in the search concluded the distress message must have been a hoax. No record was ever found of a ship with the call letters they had heard.

The vessel that had stood by for two and a half hours (the *Bougainville*) had called the Coast Guard to ascertain if there was any known distress in the area. The Coast Guard replied that they knew of none and told them they had probably just been seeing the lights of a foreign fishing vessel.

But what about the flares they had sent up? What about the distress calls they had sent while the transmitter was being cranked—hadn't any of them gotten through to the thirty or so ships estimated to have passed within a hundred and twenty miles of them?

Why had the *Texaco Oklahoma* broken in half anyhow, after having just been issued a certificate of inspection?

And whatever happened to the bow section with her captain, Richard B. Hopkins, of Camden, Maine, and the twelve men who, so far as anyone knows, made no attempt to abandon the still floating half ship in either of the lifeboats or inflatable rafts? Is it still out there, voyaging endlessly like so many other derelicts that appear, disappear, then eerily appear again?

Britain asked the United Nations to organize a seek-find-and-destroy campaign against the many "ghost ships" that have proven themselves hazards to safe navigation. Craven interest in my own skin leads me to hope I will never encounter one of these mindless wanderers, particularly if I am aboard an LNG carrier.

▶ The Greek *Herakos* was rammed in 1974 by one such ship which left a ten-foot dent in her bow. The incident occurred in the North Sea, but an immediate search by Danish Coast Guard helicopters turned up no trace of the nameless vagrant.

▶ In 1973 crew and passengers of the Italian cruise ship *Capo Verde* positively identified the *Columbian Duarte*, which had disappeared in a storm off Venezuela in 1948. Since then the *Duarte*, leprous with rust and peeling paint, has been seen nine times.

▶ The oldest of all the spectral ships is the freighter *Dunmore*. Her crew abandoned her in 1908 in mid-Atlantic when a cargo shift led them to think she was bound for the bottom. A dozen or more times she has been seen among ice floes of the North Atlantic, but with phantom perversity she eludes all air and sea searches of the International Ice Patrol.

▶ The biggest of the cruising ghost ships is the 13,000-ton *Baychimo*, stranded in pack ice and abandoned by her crew in 1931 near Point Barrow, Alaska. Although she eludes those who would destroy her as a menace to shipping, she has been sighted some sixty times in the Beaufort Sea north of Alaska's northernmost point.

How is it that some ships, untended, rust eating at their vital parts, can for years endure the flexing and bending stresses imposed by the sea's vindictive whims? Nobody knows.

How is it that others, carefully tended, well manned, can snap in two as did the *Edmund Fitzgerald,* the *Martha Ingram,* and the *Texaco Oklahoma?* Each time a U.S. ship sinks the Coast Guard publishes an exhaustive report of the accident. I have studied many of these. "Probable cause" may go on for pages, but what the educated conjectures sometimes add up to is that nobody knows.

Much more is asked of the LNG ship's hull than is asked of the hull of the conventional tanker. The LNG hull forms must support tanks which contain the frigid liquid. These tanks are constructed of many different materials from aluminum to fiberglass to cement, and each of these materials expands and contracts at its own peculiar rate when exposed to the thermal extremes imposed by liquefied natural gas. The result is a set of disproportionate pulling, twisting forces interacting simultaneously within the tank's shell which must remain intact if it is to contain LNG safely. These forces from within are constant. The hydraulic forces from without are variable. The result is a hitherto unknown combination of stresses which are of crucial importance when considering fatigue life.

My investigations indicate that more than one industry

architect feels that each of the many systems of containment should be evaluated separately. Primary bending pressures, thermal pressures, and fluctuating ballast pressures can subject the ship's hull to serious stresses which are impossible to predict even with thorough testing.

Vibration is another factor which must be considered for its own unique application to the LNG vessel. Everyone who has ever spent time at sea has experienced the effect of vibrations: running lights of an oncoming ship look like sparklers even though you bend your knees and try to stay loose so you will be in sync with the shuddering deck. Silverware, plates, and glasses stutter across the table and fall to the deck; writing is difficult and reading is blurred. I have watched delicate navigation and communications instruments vibrate at such violent levels that they are shaken to death in a comparatively short time. These vibrations are transmitted to the most remote regions of the ship hull, weakening through fatigue the critical framing members.

A conventional oil tanker when fully loaded sinks so far below the surface of the water that her vibration level, though still noticeable and annoying and damaging, is much less than that of an LNG ship. Since a full load of LNG is only half as heavy as oil, the vessel sinks only half as deep, never enough to take full advantage of the support and the shock-absorber effect afforded by deep hull submersion.

The so-called sail effect is still another problem uniquely troublesome for the LNG ship. Her high hull, even when loaded, leaves a towering wall of steel above the waterline, exposed to the forces of the wind. At sea, this results in more severe rolling. In confined waterways and congested harbors where there is little room for error in maneuvering, the sail effect is particularly undesirable.

At sea, the unleashed forces of a storm have to be experienced to be believed. Gear is stripped from the decks, welds are torn, nothing remains watertight. Salt water shorts out, corrodes, and rapidly destroys running lights, automated monitoring gear, and remote control devices. I have never been on a vessel that snapped in two, but a tanker I sailed on last year had three V-12 diesel cargo pump engines torn from their mounts and swept overboard in a storm. Joseph Conrad knew what he was talking about when he said, "I have known the sea too long to believe in its respect for common decency."

An LNG vessel's above decks are cluttered stem to stern with a maze of intricate and relatively delicate cargo-monitoring and handling-systems gear that is entirely exposed and susceptible to damage by seas which routinely break on deck during the course of any ocean voyage. To underestimate the destructive power of the sea can spell disaster. Wave force has been measured at 7,099 pounds per square foot. From trough to crest, waves of 112 feet have been measured, and in a tidal sea that distance increases to an incredible 220 feet.

Unable to ride a swell in the tradition of her smaller predecessors, an LNGC, standard size, will punch through the waves at a service speed of twenty-three knots as compared to a tanker's fourteen to fifteen knots. Living on the stern, isolated and removed from contact with the sea, one fifth of a mile away from the bow, those on the navigating bridge have no way of knowing how bad things really are forward. Severe hull damage could easily go undetected until the vessel reached her next port of call.

What will happen when the LNG ship's control devices are destroyed by those mean green seas, when her unprotected cargo piping is wrenched from her decks, or her bows stoved in from bashing head on into a wall of oncoming water? Few ports or repair yards will put out the welcome mat to a ship so potentially dangerous.

It will be remembered that, in the case of Japan's damaged nuclear ship *Mutsu*, several ports of her own country refused to accept her.

June, 1975, marked the launching of the first 120,000-cubic meter prototype LNG carrier, the *Ben Franklin*. During the next year order books swelled, and by September 1976 forty-three even larger vessels of six different tank designs were under construction or on order.

Questions concerning long-term performance, new materials, untested techniques remain unanswered.

Bigger is better? This concept gives pause to some naval architects who deplore the lack of a growth period. They would like to allow for some testing of sophisticated hardware on the smaller ships before barreling ahead with installation on the biggest afloat.

Men who have spent their lives in the shipping industry view this with concern for economic reasons if for no other, knowing the boom is speculative.

Proceed with caution?

Nobody seems to be listening.

Subcontractors lacking financial fortitude and integrity sometimes compromise on product quality unbeknownst to the shipbuilder or owner. An engineer who works for a major U.S. electronics company confided in me that the ship collision avoidance system which they market is flawed and, furthermore, the company knows it.

In the extreme heat of the Persian Gulf their equipment goes unexplainably crazy, giving wrong information. Unreliability, caused by high humidity, sand and salt in the air, voltage surges from the ship's generator, and the ever present vessel vibrations, is a problem still to be solved by this prestigious manufacturer. And yet their hardware is being aggressively marketed worldwide for inclusion in LNG ships. I stare in disbelief at their advertisements in the trade journals.

"Unforgivable." That was the word heard over and over after a rear cargo door blew off the DC-10 in what was, up until then, history's worst air disaster. The date was March

3, 1974, and the plane crashed in a pine forest north of Paris. The passengers and crew were scattered into an estimated 18,000 pieces that never could be sorted out, but the guess was that they added up to 350 persons.

Two books and many thousands of outraged words were inspired by the allegation that "everybody" knew the door was defective: the builder of the plane, the subcontractor, and the Federal Aviation Administration.

The door had already failed in one accident, though with no loss of life. For this the company got a tap-on-the-wrist reprimand. They promised to correct the defects and then, badly in need of money, sold one of the DC-10s, known to be flawed, to Turkey at a bargain price.

Much was made of the fact that Turkish safety regulations and pilot training were scandalous as compared to the more rigid standards in other countries, and shock was professed that, aware of this, McDonnell Douglas, the builder, pushed the sale.

Corporate greed being what it sometimes is, it surprises me that anyone was surprised. Economic survival is accomplished routinely by falsifying or, with a shrug, taking that calculated risk with other people's lives.

I confidently expect that U.S. shipbuilders will sell LNG vessels to any nation, regardless of its standards, so long as that nation is able to pay the price.

But when the accidents start happening to these ships, millions of words will be written and much will be made of the things that "everybody knew."

And it will be sadly remembered that the voices crying caution were not heeded in the high-decibel wilderness of ship construction.

Chapter IX

ACCIDENTS WON'T HAPPEN?

"Captain Prays after Liberian-Flag Voyage." So read the headlines in my March 1977 issue of *American Maritime Officer*. Small wonder that his safe arrival in Newport News seemed like such a miracle to the captain of the *Golden Jason* that immediately upon arrival he asked for directions to the nearest Greek Orthodox Church to thank God for safe deliverance.

It seemed like a miracle to the U. S. Coast Guard commander too.

The *Golden Jason's* boiler had failed off the North Carolina coast, and when the tanker with her nine million gallons of fuel was towed to the Virginia port it was promptly condemned and ordered out of commission. Initial inspection found her engine dead, her lifeboats inoperable, and her emergency fire pump useless.

The Coast Guard commander said, "They'd put in an extension cord and run it a half mile down the ship. I've seen some awful things on the water, but nothing like this."

One "awful thing" has come fast on the heels of another during these past several months, with 1976 copping the record for tanker accidents and 1977 off and running as a strong contender. The Armada of LNG carriers coming into view does nothing to brighten the prospect.

BLAST president Gene Cosgriff said in the 1977 memorial

service for the men killed in the TETCO disaster, "The recent rash of oil tanker and barge accidents has been environmentally devastating. Yet had one or more of the spills been LNG we would have been counting dead human bodies instead of dead fish or birds."

The weakest link in the LNG system, most seem to agree, is its carriage by ships. It has been said that the best thing the LNG shore storage tanks have going for them is that they don't move. The same thing could have been said about steam engines.

When they were first put onto boats and moved into the water, the earth's most hostile environment, safety went overboard. Prior to 1838, 230 of the 260 recorded steam engine accidents causing loss of life and substantial property damage occurred in steamboats. And most of the rest of them occurred in locomotives, which also moved. Through trial and error, however, the steam engine finally operated predictably and safely.

No time is being allowed for trial and error when LNG carriers go from the drawing board right into the water. When Lieutenant Commander H. S. Williams, U.S.C.G., was Chief of the Hazardous Materials Division he said, "New cargoes are proposed for shipment in large bulk quantities before all aspects of the transportation scheme can be evaluated. Supertanker shipments of LNG are one good example."

Experts who make their money painting rosy pictures would lull us with statistical projections such as the ones that show that the risk associated with the proposed importation of LNG to Staten Island is estimated to be about one fatality every ten million years for those living in the area.

They make cheery statements to the effect that the probability of accidents should approach zero if the rules are followed.

I am not an expert, but I have been there and I know that the rules are by no means always followed.

Furthermore, those who promise safety seem always to invoke the Coast Guard, that small branch of the service having probably the most—if not the only—unblemished image left to us, and we are told that they have, for the handling of LNG, a special set of requirements.

Let's take a look.

The Coast Guard's *Liquefied Natural Gas, Views and Practices, Policy and Safety* lists eighteen requirements. Those who are impressed may be surprised to find that all eighteen of these requirements *may* be placed upon the LNG transit operations by the local captain of the port *at his discretion.*

Also, fourteen of the eighteen have been applicable to ordinary oil tankers for years and sadly have not kept tanker accidents from escalating.

The four new regulations that do apply to LNG ships only are:

> On initial entry require the vessel to anchor and
> be inspected prior to permitting transit.

Note that the regulation addresses itself only to initial entry. Are we to believe that for the rest of the vessel's life, which may be twenty-five years or more, she need not be inspected until she reaches her dock after she is already deep within the harbor or up a river? It is when a vessel gets a few years under her belt that the equipment begins to fall apart and malfunction.

When a vessel is at the dock, hours from the open sea, is no time to find out she is leaking or that critical equipment is inoperative. But the *Descartes*, a French LNG tanker, on only her second voyage, got all the way into Boston Harbor and was offloading when it was discovered she had cracks in her tanks' membranes. The crew had disguised the leaks by purging the areas with inert gas; consequently, Coast Guard inspectors found no LNG vapors present when they tested later with monitoring devices.

In a 1973 *Washington Monthly* article about the supposedly strict safety regulations for LNG and LPG, writer Timothy H. Ingram quoted Coast Guard Lieutenant David Blomberg's comments concerning ships that had entered Boston Harbor: "Those last two LPG ships in here were shitboxes. One didn't even have the gas detector system turned on. . . . It hadn't been properly set and they were getting a continuous alarm on it, and had just shut it off entirely. They didn't have a repair manual for it, didn't know how to fix it, and couldn't have cared less."

Ingram accompanied Blomberg in a Coast Guard launch out to the Norwegian LPG carrier *Havis* where they found that the crew, rather than an international certification team, as required, had set the pressure relief valves and there was no way of knowing whether true pressure readings had been set or not. The captain signed the necessary standard form and assured the inspector that the gas detector system had been checked continuously. But when the system was turned on the gas alarm went off, red lights flashed, and the indicator showed that flammable gas was present right there with them in the control room. When the captain of the ship was told he could not unload his LPG until a shore repairman came out and fixed the defective system he was furious and was quoted as saying, "You can't run a ship like this without expecting some minor problems."

Another Coast Guard regulation unique to LNG ships would

require vessel to be escorted by U. S. Coast Guard.

The *Yuyo Maru,* it will be remembered, was being escorted at the time of her fatal collision.

For an entering LNG ship, the local Coast Guard captain of the port also may

> establish a sliding safety zone around vessel during vessel transit.

When the details of the U.S. liquefied natural gas importation began to emerge, citizen concern about the possibility of collision in the tortuous waterways in the New York City area was answered by statements like the one made in October 1973 by Admiral Bender, commandant of the Coast Guard, saying that a security zone would be established around LNG vessels while under way through which no other traffic would be allowed to pass.

Just two years later, in October 1975, when the captain of the Port of New York issued his contingency plan this regulation said that an LNG ship's passage to the dock "will be closed to *deep draft* [their emphasis] vessels during the entrance of an LNG vessel, but will not be closed to tugs, barges, and to other similar craft."

The fact, plainly apparent to anyone who has sailed this busy harbor as many times as I have, where over 25,000 cargo vessels transit a year, is that it is just plain impossible to halt traffic for a single LNG ship.

The captain of the port may, at his discretion,

> restrict vessel entry and movement to periods when there is good visibility.

On her first voyage, during rain and snow, the barge *Massachusetts* entered New York Harbor January 9, 1974, and discharged its 1.3 million gallons of liquefied natural gas at Greenpoint, Brooklyn. When queried, it was announced that the captain of the port considered it to be good weather.

Before the LNG ships, ports had never seen a speed limit.

Local LNG port contingency plans now call for one, even though its value remains a subject of heated debate.

A report, *Nature of Ship Collisions Within Ports*, was prepared for the National Maritime Research Center in April 1976, and it says, "The maximum speed is left up to the discretion of the pilot and the ultimate responsibility of the captain. This is understandable since the velocity and direction of the wind and currents are two major variables which may dictate a ship's maneuverability and control of a vessel could be jeopardized by speed restrictions."

Any experienced helmsman can tell you that the rudder becomes ineffective at slower speeds, often resulting in a total loss of steering control.

The local captain of the port's unique discretionary powers can mean that neither speed limits—if arrived at—nor any of the special LNG requirements will be enforced at all.

If I am making the situation sound chaotic, it is only because it *is*, with misunderstandings and misinterpretations at all levels. A spokesman for Distrigas, the company that began receiving LNG shipments into Boston Harbor in 1971, argued with me recently that the Coast Guard made the rules and they could not be broken. The captain of the port, he said, had no discretionary powers whatever.

What about those other fourteen regulations, the ones that have always applied to ordinary oil tankers and are being extended to include LNG ships? One of the old ones is that they shall

> require a pre-transfer conference between the ship and the terminal personnel.

This sounds good, and "conference" is the sort of word that somebody swiveling in a chair on dry land might think appropriate. Maybe an LNG pre-transfer conference will differ from the kind I've heard hundreds of times on tankers:

"Hey, what's happenin'? C'mon—give it to me, big boy."

"Okay," as the cargo pumps are started, "comin' at ya!"

The terminal personnel with whom we deal when offloading very frequently have the dock job as a number two moonlighting job. It's easy, undemanding; there are many hours when there is absolutely nothing going on in the dock shack so that they can catch up on sleep, read, play poker, whatever, to the point where we are made to feel as if our needs for their attention are an imposition.

Another old/new regulation requires

effective communications during cargo transfer.

I remember a time in Fort Lauderdale—it happened to be April Fool's Day—when I was discharging fifteen million gallons of gasoline from tank ship to tank farm. The tank farm was a great distance from the ship, as they often are, but with a permanent telephone line between the plant and the dock with someone standing by on each end.

We were pumping the usual half million gallons per hour to the shore tanks with the understanding that when the shore tanks were full the plant supervisor would need only to call the dock man, who would signal me to stop the ship's cargo pumps.

The shore tanks filled; the supervisor called the dock to pass the word; busy signal. If he shut his valves against the ship's pumps the rubber ship-to-shore cargo hoses would burst and we would be pumping into Fort Lauderdale Harbor and onto surrounding beaches.

The gasoline started pouring out over the top of the shore tank. The supervisor jumped into his truck to race for the ship, but losing control on a corner, he rolled the pickup into a ditch. Bruised and bleeding, he continued on foot to the ship where, minutes later, he crawled to the top of the gangway gasping, "Stop—stop the pumps."

I yelled the order, followed him down the ladder, and the two of us stormed into the dock shack. The man was still busy on the phone and appeared to be writing down a message. As we entered he did not look up to see who it was, just held up his hand like a cop stopping traffic, indicating he was not to be interrupted. "Okay," he said into the phone. "Okay . . . orange marmalade . . . need anything else?"

Another time, loading chemicals in Puerto Rico, I had put the shore personnel on standby to shut down the pumps and valves when the cyclohexane reached the top of the ship's tank. Everything was set. The man was standing by his valve to stop the flow from shore the moment we let him know the tank was full. We didn't have the usual walkie-talkie between ship and dock here so we just had to scream back and forth.

I had my face in the tank watching the level rise and trying to breathe although the fumes were being exhausted right at me. I yelled, "Okay, shut it down!" It kept coming up at me. I looked around. No one was on the dock. The man had gone. A geyser of cyclohexane burst out of the tank, pouring onto the deck, splashing onto the crew, and running over the side into the harbor.

What had happened? The dock man had spotted his relief man coming way down the pier. Anxious to get off work, he had left his station.

If I told stories like this all day—and I could—they would never affect the scientists' statistical evaluation of risk. But one good big accident *would*. And these are the ways in which those good big accidents happen. Crewmen have respect for any dangerous cargo at first, but fear is replaced eventually by carelessness. Nothing happens the first hundred times you handle it and so you relax . . . nothing will happen—at least not to you.

As the marine consultant I quoted in Chapter IV said in discussion of the accident he feels to be inevitable with LNG: "Many will become complacent."

Another long-standing oil tanker regulation that has been extended to LNG vessels says to

> prohibit welding, burning, hot work, smoking, open lights, etc. . . . while the vessel is moored at the terminal.

Twelve men were lost when the Panamanian tanker *Claude Conway* exploded and sank off the coast of North Carolina on March 20, 1977. A welding torch caused that explosion.

More than once I have been on ships where welding was done on a deck that was not gas free. Everybody knows that it is dangerous. It is against regulations. But it is done. From what I have seen, it will be done on LNG ships.

And smokers will smoke. Cigarette lighters were found under the rubble following the TETCO accident on Staten Island. The regulation prohibiting smoking is one of the hardest to enforce. I remember an ordinary seaman who was initially scared of working on a tanker carrying high octane gasoline. We had completed only two voyages when I caught him ducking behind a pump house to "grab just a quick smoke"—while we were offloading.

Clyde T. Lusk, Jr., a U. S. Coast Guard captain, said not long ago, "A marine casualty is a hard-to-disguise failure. Many occur because of malfunctioning equipment, violation of the law, or errors in judgment; all are manifestations of non-perfect safety systems. Some in the safety business find it necessary to periodically justify their programs by reciting (or should I say 'juggling'?) statistics to show an improved safety record."

The justifications and the juggling of statistics are usually done, not surprisingly, by those who work for the industry or otherwise have a vested interest. Lee Joseph, who, as previously mentioned, has opposed siting of LNG tanks in the Camden-Philadelphia area, said bluntly when we discussed this matter that relying on industry spokesmen is like relying on the fox in charge of the hen house.

The experts who would placate our fears can hardly these days suggest that the safety record has improved and so they let statistics roll trippingly off the tongue to prove that the probability of LNG accidents is so remote that only the hysteria-prone few would worry.

We are encouraged to believe that because the Coast Guard has promulgated regulations specifically for LNG ships in port it is they who have put the final stamp of approval on the importing of LNG and its siting.

The implication is that the Coast Guard wouldn't let us do it if they didn't think it was safe.

The fact is that the Coast Guard inherited that decision from the Federal Power Commission. The Coast Guard is thus obligated under the law to formulate requirements to accommodate that decision.

To promulgate is one thing. To enforce is quite another.

The Coast Guard is spread too thin to do this job effectively. Too much is asked of the 36,000 men and women who, understaffed, underfunded, must respond to over 40,000 distress calls annually, operate and maintain all their vessels, run icebreakers in northern waters, track icebergs and weather patterns, run electronic navigation systems, provide emergency medical aid to crews of all vessels at sea, maintain thousands of lighthouses and marker buoys, run the Coast Guard Academy and many other training facilities, handle marine inspection, conduct licensing examinations, track down all kinds of smugglers, enforce federal laws on the high seas, effect search and rescue, and in addition to all these and other tasks not mentioned, now take on the enforcement of the new 200-mile fishing limitation along our coasts.

Coast Guard regulations, especially as they apply to foreign-flag ships entering U.S. ports, often go unenforced for another reason.

After the *Argo Merchant* sinking, Admiral William O. Siler, commandant of the Coast Guard, told the Senate Commerce Committee that his agency did not enforce safety standards on Liberian- and other foreign-flag ships as it had been empowered to do by the Ports and Waterways Act of 1972 because he did not want to cause trouble abroad.

If the Coast Guard were to clamp down full force on every foreign-flag offender there is always risk of retaliation when U.S. ships visit foreign ports. Foreign policy can dictate whether or not the Coast Guard should look the other way when a foreign flag is not obeying every regulation.

How far is the Coast Guard prepared to bend to prevent "trouble abroad"? Dare we step on the toes of a foreign government even though one of their ships carries enough LNG to level one of our cities?

This is not my domain and I am glad. I see enough right under my nose to worry about. There is, for instance, the power struggle that sometimes takes place on the navigating bridge between pilot and captain.

When a tanker, LNG or otherwise, arrives at the outer limits of a U.S. harbor the law requires that she take a licensed pilot on board to advise the captain concerning local conditions and particular hazards to navigation.

Once aboard, the pilot customarily takes over the navigation and maneuvering of the vessel, calls the shots, even though the captain—or, in his absence, the mate—can take it away any time he sees fit.

Consequently, what you have during this most critical phase of a ship's operation is a disturbing dichotomy: the licensed pilot, expected to handle the ship in close quarters in waters he knows like the back of his hand, and the ship's captain looking over his shoulder, always in command, one hundred per cent responsible for the ship.

At first glance it might seem that this captain/pilot relationship would balance each other's strengths and weaknesses. But differences of opinion can result. Ego and professional pride come into play.

The captain often has difficulty believing the pilot can have a feel for the unique handling characteristics of *his* ship since the pilot is off and on so many other different ships so fast.

The pilot, on the other hand, views himself as a specialist since he maneuvers ships in close quarters every day and knows this waterway better than the captain can possibly know it.

How many times have I seen the man at the wheel look back and forth frantically between captain and pilot, confused because each has given a different rudder order or engine speed or direction command. When this happens within the constraints of a bend in a river, a traffic meeting situation, or during docking maneuvers the results can be serious.

Each port has its own pilots' association and some ports have several in competition with each other. An applicant must be certified as an able-bodied seaman, which is just one step up from the lowest ship-board job. He must also pass a Coast Guard examination to show knowledge of the area for which he seeks pilotage. The law requires no demonstration of ship-handling ability. Nor is any simulator training required.

I doubt that I would board a plane if I were told that the pilot had experience as an aviation mechanic but had never actually flown the plane before. There is no chance that I would climb aboard if I were told that there might be two

people in the cockpit fighting over the controls during critical maneuvers.

Acceptance into a pilots' association is often political.
Some associations require large amounts of capital up front
before an application will be accepted. Apparently the applicant must be male; I know of no women pilots.

We have no rules that stipulate how long a pilot may
work. From the Southwest Pass at the mouth of the Mississippi to Baton Rouge a pilot may be required to maneuver
without rest for twenty hours straight, assuming there are no
delays. On shorter transits such as the Houston Ship Channel or waterways around New York City, pilots will often
ride one ship out and another back and another out and so
on.

All mariners have to endure long periods of work. Early in
my career I sailed as third mate worldwide on one ship for
387 consecutive working days, fifteen to eighteen hours a
day, and I well remember that my efficiency curve de-
scended as stress and exhaustion took their toll. After that
length of time I was unfit to be in charge on the navigating
bridge.

It is very possible that eventually special rules will be in
force stipulating that the pilot who comes aboard an LNGC
may work only a reasonable length of time at one stretch. I
hope this will be true. But what I do know for sure is that
LNG vessels will be encountering other vessels which are in
charge of exhausted pilots and captains who have worked
long past peak efficiency.

The language barrier presents problems that further com-
plicate a pilot's job. Very often no English is spoken at all by
captain or crew, requirements notwithstanding, and if a little is spoken it often vanishes in tight situations when it is
most needed. The English language is full of inconsistencies.
I once witnessed a near accident when a pilot ordered,
"Left to fifteen degrees." The "to fifteen" order designates a
rudder angle relative to the center line of the ship and the

order as interpreted, "two fifteen" rudder, nearly piled us into a bridge abutment.

To say that we often have kooks as pilots is only to say that pilots are people. We took one such kook aboard our loaded tanker headed up the Delaware River. As mate on watch I was working the throttles as per pilot's instructions, watching to make sure the helmsman carried out his rudder orders correctly. The captain had been below in his cabin for about an hour and everything was routine until we entered a straight stretch in the river and the pilot ordered full speed ahead.

"Full ahead," I repeated, easing the throttle levers forward.

"Hard right," the pilot ordered.

The helmsman responded, "Hard right," and the spokes blurred as he spun the wheel to starboard.

Why hard right at this beginning of a long straight stretch? "Pilot, what's going on?" I asked.

The tanker dove violently to the right, heading out of the channel into shoal waters. At the last possible moment the pilot yelled, "Hard left!" to break the turn and regain control of the ship. Just in time, this was done and we settled down on our original course.

I edged closer and asked, "What happened there?"

"Son, I've been doing this run for so many years that it's like driving a bus. Throwing this thing into a near miss is the only way I can get the juices going."

If the captain had been on the bridge, the pilot probably would have found a less dangerous way to "get his juices going."

Make no mistake, at sea the captain is God. His word is law. He can marry you, bury you. He is judge and jury. A captain loses his responsibility aboard ship at only two times: when the forefoot of the bow crosses the entrance of the graving dock where he is assumed not to be experienced enough to carry out the delicate operation of putting the vessel in drydock, and when the vessel is in transit of the Panama Canal.

Canal pilots are probably the best trained and most competent in the world. The canal is too important to international commerce and national defense to have some inexperienced, possibly fruitcake captain go piling into the locks and tie this waterway up for any length of time, no matter how short.

I remember an incident going eastbound through the canal when the captain, reluctant to relinquish authority, continued to roar his commands, countermanding the pilot's instructions. The pilot put the engine on stop and said quietly, "Mister, you have three options: we can anchor right now and you can sit here until hell freezes over, or we can call for the tugs and have this vessel taken through as a dead ship, or we can put you ashore and you can ride a train from Panama City to Cristobal through the Zone and meet us on the other side. The choice is yours."

The captain, his face purple, stalked off the bridge.

Two or three years ago, knowing that U.S. pilots had never handled any carrier as large as an LNG, I started asking some of them how they felt about taking the conn on one of these ships. Their answers ranged from a blasé "No sweat—when you've seen one ship you've seen 'em all" to a fatalistic "When you gotta go, you gotta go."

Most pilots agree in their criticism of the Vessel Traffic System (VTS) which been inaugurated in a few U.S. ports and will eventually be in all of them. "All we need is somebody else telling us how to do our job," they say.

Communication on radiotelephone is apt to be so garbled with overlapping talk that nobody can tell who is talking to whom. CB-ers, comparatively, have a private line. Even Channel 16, the international distress channel, is jammed with inconsequential jabbering that seriously interferes with urgent communications.

The radar and television cameras that are set up, sophisticated though they are, cannot effectively monitor all the waterway system of the harbor, which may extend fifty miles or more. Although VTS is a step in the right direction, no one at this point is saying that it is working perfectly. Since participation is voluntary, many vessels will continue to move without the Coast Guard's knowledge.

My oldest brother, Bruce, a captain in the Merchant Marine, told me of having visited a VTS control room that had a huge scale model of the river harbor with all the bends and tributaries and docks and installations. When information was received from such ships as chose to participate, 3 × 5 file cards with the names of the ships and their vital statistics were put in place and moved along as their progress was radioed in. He happened to be there when a gust through a suddenly opened window blew all the cards onto the floor. Nobody could remember where they belonged.

I have pamphlets detailing the benefits of the sophisticated system and as I looked through them I was remembering that my brother also told me that when maritime personnel stop by for a look at that facility it is not always to observe harbor traffic. It is well known that up and down the river on a clear warm day you can zoom in on what has come to be known as "Nudie Beach" and on another secluded spot that has local fame as a lovers' lane. Who could say he wouldn't take an occasional look just to rest his eyes from the boring job of monitoring harbor traffic?

The rules and regulations that I talked about earlier in this chapter have been port requirements that do not extend beyond the geographical limits of the port. *Rules of the Road* are international, agreed upon—no small miracle—by all the maritime nations of the world and recently updated with the guidance of IMCO (Intergovernmental Maritime Consultative Organization), which is a sort of United Nations of the sea.

Even the *International Rules of the Road* are disobeyed. One of the most important of these says:

> Every vessel when on the water shall in fog, mist, falling snow, heavy rainstorms or any other condition similarly restricting visibility, go at a moderate speed having careful regard to the existing circumstances and conditions.

I have never been on a ship where we slowed down a single knot in zero visibility at sea. A third mate of my acquaintance was even fired for objecting to full speed in dense fog.

Alvin Moscow in *Collision Course* said, "Shipmasters facing the choice between delivering their passengers and/or cargo to port on schedule or slowing down for safety's sake in fog have consistently chosen to risk traveling at top or near top speed through fog. On-time arrivals have always been a measure of a captain's ability. Anyone could stop in a fog and wait for a day, two days, or a full week, losing money for the company all the time incurring the wrath of passengers and shippers."

When I was a midshipman on the S.S. *Brasil* on what was billed as the "Northlands Cruise," we sailed from New York to Honningsvag, Norway, the northernmost tip of Europe, via Iceland, then on down through the Norwegian fiords and into the Baltic before crossing the North Atlantic on her return to New York.

We were almost constantly surrounded by fog, typical in this iceberg-infested area. Ten thousand icebergs "calve" every year, break off from the mother glaciers in Greenland and, with summer warmth, move down into these waters. Some are enormous; one monster is said to be the size of Connecticut. As we blasted along at about twenty-five knots the cruise director took over the P.A. system to announce solemnly to the passengers, who seemed oblivious to risks involved, that we were now passing close to the spot where, every April 14, the Ice Patrol drops a ceremonial wreath in memory of the sinking of the *Titanic*. No ship, he added, has since been lost due to collision with an iceberg in the area covered by the Ice Patrol. The passengers could also take comfort from the knowledge that we were protected by radar.

I had at that time just learned the *Rules of the Road* and these words were firmly fixed in my mind: "It must be remembered that small vessels, icebergs, and similar floating objects may not be detected by radar."

The only rule of the road being observed was: "prolonged blasts of the ship's whistle at intervals not to exceed two minutes." And this was discontinued when one of the passengers complained that the noise was making her chihuahua so nervous it could not sleep.

Perhaps the most important rule of the road in terms of safety has to do with keeping a sharp lookout. An international conference on modernization of the rules gave it elevated status by devoting an entire rule to that subject:

> Every vessel shall at all times maintain a proper look-out by sight and hearing as well as by all available means appropriate in the prevailing circumstances and conditions so as to make a full appraisal of the situation and of the risk of collision.

"Proper lookout" on one vessel which I saw consisted of a dog chained out on the wing of the bridge.

I had spotted the ship about three miles away, and when I picked up the binoculars to get a closer look I saw the dog. It was obvious from his behavior that he began barking furiously when he sighted our ship. It was like watching a silent movie. I saw him jumping up on his hind legs, front paws resting on the railing, and as the dog's jaws went up and down I could see a man down on the main deck look up and run aft and up the outside ladders to the bridge, pick up his binoculars, and look at me.

As one shipmate remarked, the rule doesn't say who has to keep a proper lookout, only that one must be kept.

That ship at least had a lookout, "improper" though it was. Some ships are not even that careful. The undersea explorer Jacques Cousteau recently told a congressional group of having been almost run down more than once in the summer of 1976 by foreign-flag ships when he and his team were carrying on their explorations near Greece. They had to set off flares and fire blank pistols at the ships to wake up helmsmen who seemed to be asleep at the wheel.

Disregard for the *International Rules of the Road* is usually based on ignorance, ignorance of the rules themselves and of the awesome cargoes being transported these days whose lethal possibilities go far beyond the carrier itself and her crew.

The vast majority of tankers since the first one, the *Glückauf*, nearly a hundred years ago, have carried oil. With the development of superior tank coatings, shipowners realized that they could carry any liquid in bulk. Texaco has hauled fresh water from the Gulf of Mexico to Trinidad. Almost every salable liquid has been transported—milk, paint, orange juice, molasses, castor oil, wine and, in the last dozen or so years, chemicals.

Some tankers now carry so many different kinds of chemicals that they are referred to as "drugstore ships." With defoliants, poisons, corrosives, and acids of every conceivable type being loaded in large quantities into ships' tanks, the U. S. Coast Guard, although it has no power internationally other than to recommend or express objections, wrote to the Intergovernmental Maritime Consultative Organization:

"The United States wishes to express its views on the

problem of the sea transport of dangerous chemicals in bulk, so that delegates may have the opportunity to reflect on what action may be undertaken by IMCO. We consider certain chemicals to be highly dangerous. This is often magnified by containing the chemical in other than its natural state by artificial means such as low temperature or high pressure. We believe that the chemical carrying ship, which is actually a gigantic package of such dangerous cargo, presents a port-and-population hazard similar in nature to that of a nuclear vessel. However, there is a lack of any formal recognition for this type of vessel."

Concern was expressed about products with "grave toxic and/or explosive characteristics" nearly all of which were carried by vessels of other than U.S. registry which would be trading in and out of U.S. ports.

My youngest brother Derek works as an able-bodied seaman on the largest chemical carrier in the world, the 42,000-ton *Seabulk Magnachem*. Although primarily involved in the U.S. coastwise transportation of caustic soda, the *Magnachem* may carry as many as twenty-seven different products.

On his last visit home, my brother and I sat down to read the listing sheet of possible chemical bulk liquids his vessel may carry. It was amusing, initially at least, to try to read and pronounce their names: aminoethylethanolamine, tetrahydronaphthalene, and one that had sixty letters—we counted them.

I asked him how much the crew understood about these chemicals. Not much, he said, they just knew they were mean stuff.

More explicit was a detailing of the hazards of some of those chemicals which appeared in a British publication, *New Scientist:*

"Perchloroethylene vapour causes dizziness, nausea, and vomiting; in high concentration, stupor. Liquid irritates eyes.

"Toluene. Highly flammable. . . . Vapour produces dizziness, headache, nausea, and mental confusion. Also irritates eyes and mucous membranes. Absorption through skin causes poisoning and dermatitis.

"Benzene. Highly flammable. . . . High concentrations of vapour cause dizziness, headache, excitement, and unconsciousness. Repeated inhalation of low concentrations over a considerable period causes severe blood disease, including leukemia.

"Vinyl acetate. Highly flammable. . . . High concentrations of vapour may be narcotic. Liquid irritates eyes and may irritate skin by its de-fatting action."

Cyclohexane was also included as being highly flammable, irritating to eyes and respiratory system. I can attest to the accuracy of this description. When I joined a ship loading this chemical in Puerto Rico I took over my first watch and looked down into the tank in the usual way to see the rising level of the liquid. Nobody had told me it wasn't like gasoline.

Temporary blindness was instantaneous; my lungs were seared; I fell to my knees in a state of disorientation. The crew and officers from a previous watch popped out from behind the pump house where they had been hiding to watch the fun, laughing uncontrollably, having anticipated what would happen to a person unaccustomed to handling this type of chemical.

It's not all fun and games. Some of this stuff can kill you. The Coast Guard Proceedings of the Marine Safety Council reads:

"Vinyl chloride monomer is one of the most important chemicals in the plastics industry. In 1973 over 338 million pounds of vinyl chloride were exported from this country by water. . . . Vinyl chloride is reported to cause angiosarcoma of the liver—an almost always fatal form of cancer."

But I submit that the most corrosive, the most lethal of these chemicals cannot compare to the potential dangers of an LNG cargo.

Chapter X

SABOTAGE

Three gunmen demanding a ten-million-dollar ransom forced the pilot of a Southern Airways plane to circle the nuclear facilities in Oak Ridge, Tennessee, threatening to crash into it with thirty-one passengers aboard if demands were not met. Thousands of miles later, after nine landings, the gunmen gave up, and with tires shot up and the copilot wounded the plane was landed on foamed tarmac in Cuba.

Press accounts of threats made by two women brought the Manson cult into the headlines again when these women announced they had threatened seventy-five executives of several large industries, among them Pacific Gas and Electric, owner of Western LNG Terminal Company. The accusation was that they had polluted the environment, poisoned the water, killed wildlife, cut down trees, and falsely advertised. Those named were informed that they had been "sentenced" by the International People's Court of Retribution, and told that they and their wives would be "butchered in their bedrooms because they were living off the blood of little people."

Church bells pealed in the early morning hours of March 10, 1977, joyfully announcing that Washington's forty-hour siege of terror had ended. Twelve members of the Hanafi

Muslims had been threatening death to 134 hostages, not because of particular malice toward any of them, but because they were there, fate having placed them in the wrong place at the right time. One man was killed and twelve were wounded before this "holy war" ended. Hamaas Abdul Kallis, leader of the sect, began life in Gary, Indiana, as Ernest Timothy McGhee. One of his goals was revenge for 1973 slayings, four of them his children and one a grand-child. The life sentence meted out to the killers had not satisfied him.

One week before the Washington takeover, a 661-page federal task report on disorders and terrorism said the state of many big cities is "more desperate than it was during the riots of the 1960's." It warned of a possible increase of con-ventional terrorism and recommended that the hands of the FBI and the CIA be untied so that they could once again be able to make covert investigations, enter and search without a warrant, and hold suspects without arrest. The Hanafi Muslim files, which might have proved useful at the time of the Washington siege, had been shredded three years be-fore.

It is well known that organized terrorist groups have training facilities in many parts of the world. In order to assess the difficulties such groups might encounter in trying to come up with such powerful leverage as the atomic bomb, John Aristotle Phillips, a self-styled "strictly average student," undertook a research project and wrote a paper when he was a junior at Princeton. Its lengthy title bears the message: "The Fundamentals of Atomic Bomb Design—An Assessment of the Problems and Possibilities Confronting a Terrorist Group of Non-Nuclear Nations Attempting to De-sign a Crude Pu-239 Fission Bomb."

He used only unclassified material. His four-and-a-half-month project resulted in a design for the bomb and an esti-mate that it could level a large part of Manhattan for

$2,000, excluding the cost of plutonium. Requests for copies of the paper came from several countries; parts of it are available, but with critical deletions. Actually, obtaining the plutonium is the hard part—it is expensive, available only to the qualified few. Since stealing it is the only option for terrorist groups, Phillips hopes his paper will result in the increased safeguarding of plutonium supplies.

Terrorism has many different motivations. Some of them are as absurd as the one given by a drunken young American who held an open razor to the throat of a Japanese stewardess, threatening to take the U.S.-bound 747 to Moscow because "the United States has too many food additives."

Altruism, or that which masks itself as such, may seek to right real or fancied wrongs which can be religious, political, economic, sociologic, or environmental. An editorial in *National Review* gave credit to Keith Mano for having pointed out in *The Bridge* that "The ecological consciousness has its dark potential and in its extreme manifestation can be antihuman, the rhetorical 'air and water' serving as a mask for murderous hatred."

Usually the terrorist is young. He may be well educated, clever, or stupid. Often he is psychotic, obsessed with having been wronged, ignored, or considered by others to be a loser. Sometimes he embraces a cause in which he can see himself as a savior. For his day of glory he is willing to die or at least risk dying. Now his name will be in the papers. Now his face will be on TV. Now everybody will know who he is.

And then again a terrorist may be merely a disgruntled employee who is willing to furnish necessary information to those who can mastermind the sabotage. There are indications that some of the harm done to nuclear plants—cut wires, damage to valves and switches—was with the assistance of insiders.

In May 1976 the Nuclear Regulatory Commission (NRC) issued a security alert to all of the nation's nuclear plants.

The credentials of two NRC inspectors had been stolen and it was feared that entry might be attempted at one of the fifty-eight facilities.

The alert message issued by the commission said: "We have obtained information, not fully verified, from the intelligence community that two groups may have plans to take over or occupy one or more nuclear power plants on Memorial Day weekend or to take other actions in early June."

Nothing happened. But this did not mean that the NRC had been unnecessarily paranoid. Eighteen threats had been received previously that year.

Michael Flood, a British chemist in the Department of War Studies at Kings College, University of London, believes that it is inevitable that terrorists will zero in on nuclear installations. Indeed, there have been, worldwide, many failed attempts that could have been catastrophic.

But LNG facilities are much more vulnerable than nuclear installations. There sit the tanks in plain sight with security so lax as to be laughable if it were not so serious. Almost anybody can take a guided tour. Anybody, for the price of a stamp, can write to any one of the several facility owners and get an environmental impact statement with drawings and specifications for the whole installation.

Looking at those specifications which I sent for, I see that the domes of the tanks are comparatively thin, more vulnerable than the walls, which are usually of steel, but no more than half an inch thick. The walls of the Rossville tanks have ten feet of solid reinforced concrete. It is anybody's guess as to how much damage could be done by an employee with specialized knowledge of, let us say, an anti-tank weapon such as a shaped charge, which has a vicious potential for demolishing reinforced concrete.

A failed threat often becomes much more than a mere nuisance. In New York I once saw crowds pour out of an office building when it was evacuated because of a bomb threat that turned out to be a hoax. One woman was tram-

pled to death and several were injured. When the Southern
Airways plane circled the comparatively small town of Oak
Ridge, threatening to crash into the nuclear facility, air-to-
ground conversations broadcast by the media caused panic.
Traffic was jammed. Everybody was trying to get out of
town. One official said, "The whole town was in an uproar.
We could have handled an orderly evacuation." Oak Ridge
is situated in open country. New York City is not. There is
nowhere to go. It would be impossible to handle even an
orderly evacuation.

It is the LNG vessels themselves that provide the most
vulnerable target. An LNG travels unarmed, unprotected.
She is easily boarded from the time the Jacob's ladder is
hung over the side to pick up the pilot. In spite of No Visitor
signs—as on all tankers—supply men, repair men, dock personnel come aboard without showing credentials of any
kind, a far cry from the tight security I remember when I
was a cadet aboard the nuclear ship *Savannah:* I had to
have a special pass; an armed guard was on duty at the
gangway; the control room was locked.

Since the days of piracy, commandeering a large ship has
been rare. But according to subcommittee hearings before
the House of Representatives, 1974, literally hundreds of
crews, passengers, and private boats have disappeared without a trace since 1971, with hijacking suspected. "None of
the missing owners have been found and law enforcement
officials assume most or all have been murdered."

The Coast Guard cautioned yachtsmen: "Be aware of
becoming a target of the modern day pirate or hijacker, not
by way of boarding by force, à la Captain Teach, but by
way of stealth and trickery. . . . Know your crew! Particularly hired crewmen, but do not overlook that charming
tagalong guest you met around the marina or the dock who
was agreeable to making the voyage just for the fun of
it! . . . Check for stowaways."

Commandeering an LNG vessel would be the "easiest

pickin' this side of bank robbery." And if terrorism may someday provide us with a surrogate war, as has been suggested, this takeover would need no army, just a half dozen—or even fewer—cool, ruthless men, chosen for their know-how and their loyalty to, let's say, some nation that feels it has been economically victimized.

Assume, for a moment, that initially just one terrorist managed to sign on as a crewman on an LNG vessel. Not quite the "charming tagalong guest" envisioned by the Coast Guard, but an able seaman with faked credentials. He has come aboard where the LNG was loaded, perhaps Arzew, Algeria. A couple of days from port, New York, a lifeboat is sighted. According to international law and all rules of seamanship, it is obligatory to pick up survivors— indeed, there are penalties for not doing so. It is not hard to imagine the "survivors" receiving VIP treatment for the rest of the way into port, nor is it hard to imagine the takeover of crew and ship when the time comes. I can think of nothing to prevent it. Now they hold the knife at the throat of New York. The day of glory has arrived for these fanatical representatives of this little country which feels it has been under the heel of the world for too long.

Is all this providing a how-to, furnishing a game plan, putting new ideas into mixed-up heads? That is a top-of-the-mind question that will be asked only by those who are unaware of the sophistication, the organization, the techniques of those who started thinking along these lines many years ago.

The terrorists have such a head start that we may not be able to stop them at every turn, but we can prepare by informing ourselves. Public surveillance can work wonders. David B. Tinnin, author of *The Hit Team*, urged viewers of the "Today" show to be alert, write letters, bring influence to bear upon industry and government, forcing them to take safeguards and exercise all possible vigilance.

Although a company spokesman admitted that fear of sabotage was the reason for painting out the big red letters identifying LNG on the trucks moving through New York City during the cold winter of 1976–77, the LNG industry has little to say about sabotage.

You can read hundreds of pages of environmental impact studies, flip page after page of risk analyses with, except for a bare mention, a rather curious omission of this very important danger. With Science Applications, Inc., giving the chance of an earthquake destroying the Los Angeles LNG facilities at a comforting one in one hundred trillion years, how is it that we are not similarly soothed by some such assessment of the risk of sabotage?

This is the reason:

"The threat of sabotage may well represent the largest risk to the public from LNG facilities."

That statement was made, not by me, but by Drs. Andrew J. Van Horn and Richard Wilson of the Energy and Environmental Policy Center, Harvard University, in their report issued jointly in November 1976.

Wilson said in a report he made a year earlier, "I visited a small tank eighteen months ago and was able to get close to it, look around, and decide which valves to open to let out the gas. Are not these tanks especially liable to sabotage? And if so, why was I able to get close without a guard? A Bangalore torpedo could destroy the tank, and the dike could be destroyed at the same time. Why is not this a likely event we should guard against—especially since the advent of terrorist groups of the last five years?"

At the time of the Federal Power Commission hearings reported in Chapter II, the court was shown TV instruments purported to monitor the safety of the systems at the Rossville installations. Davidlee von Ludwig said then, "The movement of men on the scene being minded by TV is indetectable. If anyone is in fact happening to watch the screen, placed well out of any normal line of observation high on the wall of the control room, any change in condition which the TV screen might display would have become massive in size before it became noticeable on the small screen."

In further reference to the TV scanning system he said, "It would be useless to forestall entry of teams bent on sabotage. The guard fence is worthless in these respects and the entire system would be wide open to sabotage from the bay front."

In the fall of 1976, well into the writing of this book, I decided to get out and walk the boundaries of as many LNG storage sites as possible, whether proposed, under construction, or already in operation. I wanted to talk to people and ask a lot of questions. It seemed logical to begin with the northernmost facility on the East Coast and then work my way down and around the Gulf Coast and ultimately end up on the Pacific Northwest.

The Distrigas tanks stood out pure white against the Boston skyline when I arrived on a clear October morning. I had made no appointments at this Everett facility, hoping that people would talk to me more candidly than if I announced my name and mission in advance. At the small office adjacent to the tanks, I learned that the Distrigas officials had their headquarters on the other side of the Mystic River in downtown Boston.

Nobody here had time to talk to me or answer many of my questions. Not wanting to have made this part of the trip for nothing, it seemed that I might as well wander around on my own and see what I could see until somebody stopped me.

I walked around the office shack and across the railroad tracks toward the fence surrounding the tanks. It was only about three feet high in some places and looked like an easy jump, too easy to be real. Holding my camera in one hand, I put out my other hand cautiously, thinking there must be an electric eye to monitor the fence.

Nothing happened. No buzzing, no alarm bells, nobody rushing to haul me off to jail. There may have been No Trespassing signs, but I didn't see any. Yet there on the other side of the fence were the big LNG trucks loading, with the towering storage tanks just beyond.

What if they had sentry dogs? But that hardly seemed likely with the workmen loading the trucks and work crews in process of assembling a huge scaffolding against a tank wall.

What the hell. Over the fence I went, easily enough except that in the attempt to protect my upheld camera I lost my balance when I landed and managed to snag the knee of my pants.

Looking at the pictures I took that morning reminds me of the casual interest I attracted: "Hi. . . . How ya doin'?" I snapped pictures fast at first, certain that I was being monitored and that it was just a matter of time before somebody came to stop me and probably confiscate my camera.

I wandered around for about half an hour and left, as I had entered, without being questioned. This place had been in operation since 1971, the most active of all our LNG facilities. Was it possible that this was all there was to the tight security I had read about?

The next morning I went to the Distrigas offices in downtown Boston. One of the executives gave me some time. He told me at length about the merits of LNG, which I was already convinced of, and the shortsightedness of those who opposed the filling of the tanks in Rossville.

I took the bull by the horns. "What if a bunch of radicals —terrorists—were to try to commandeer the facility I saw yesterday?"

"No way. We own the road. We can close it at any time. Our security is tight, very tight."

I was nodding. "I went into the office. Everybody was pretty busy. And I assumed that, without a pass, I—"

"You assumed right," he said warmly. "You didn't know

it, but all the time you were there you were, so to speak, ha-ha, on camera."

"Really?"

"Oh yes. Round-the-clock surveillance. Hopefully, such measures are not really necessary. But with a facility like ours you just can't take any chances."

I looked down at the snag in my pants that I had gotten yesterday when I went over the fence. "Right," I said. "Oh, right. . . ."

Chapter XI

DON'T GIVE UP THE SHIP

Richard Wilson of Harvard University said, "There were mistakes in the Cleveland disaster; the tank should not have been made of brittle material and there should have been a dike around the tank to contain the spill. Also, in the fire at a Staten Island LNG tank, the insulation need not have been of inflammable material. What guarantee have we that there is not another mistake that we have not yet found? The previous ones are hardly a good recommendation for the quality of engineering in the industry."

Edward Teller, father of the H-bomb, said as recently as 1976, "Time and money spent on the safety of liquid natural gas is less than one per cent of nuclear reactors. I am suggesting that very greatly increased attention be paid to safety studies."

I say that this is one hell of a time to be forced to acknowledge the truth of such commentary. I say this is the eleventh hour. I say I see too many signs that industry in too many instances, instead of acknowledging errors, seems determined to go barreling ahead making blunders that are a disgrace to industrial engineering and a highhanded disregard of environmental considerations and human rights.

Look at Canvey Island.

Little Canvey, in the estuary of the Thames, has a lot in common with Staten Island. So far as I know, neither has a

Hilton; those who chose to live there were not looking for the most fashionable resort but for an affordable place where they could get away from it all and smell the sea.

In the summer people still go to Canvey to get out of London, forty miles away, and the population of 30,000 goes up to nearly twice that much. But when the wind is wrong the smell of sulphur, like rotten eggs, effaces the smell of the sea. Black particles swirl and settle and the beaches are choking with oily bilge. No wonder—forty per cent of this tiny island, four and a half miles long by two miles wide, is occupied by petrochemical and allied industries.

When British Gas Corporation was looking for a place to site LNG storage, Canvey Island was chosen—as was Staten Island—for its convenience. Population density was apparently disregarded, and if there were those who objected to the scheme of digging pits to hold LNG I don't doubt that they were called hysterics. I don't doubt that they were told what a great thing it was for the island and how much employment it would bring.

It is a matter of record that when the tanks were first built on Canvey they were hailed as "an idea at the forefront of technology, a unique achievement."

This "forefront of technology" idea was accomplished by freezing a ring of soil and lifting out a cylinder of earth, 187 feet across and almost as deep. Four such pits were excavated. The designers did not know that the project was to reveal how little they understood "ice creep" and the problems of frozen earth. From the beginning, nothing worked. The walls developed cracks which allowed heat to come in and LNG to leak into the ground, threatening foundations of surrounding structures.

These "unique achievements" have been built in twelve other places around the world, including three in the United States. Canvey is the eleventh one to fail. Reportedly, gas is leaking out of the ground, but the company denies the pits are unsafe, pooh-poohing hazards. In statements to the public, emotive words like "cracks" and "explosions" are avoided.

A British trade journal reports: "With Islanders already paranoid over the explosion threat from the four refineries in their back gardens, they are unlikely to give the pits the benefit of any safety doubt. As long as rumours are met with outright denials and secrecy the corporation will appear to have something to hide."

It has been estimated that when the pits are emptied they will hold up about nine months before they collapse. They must be kept in use if they are to be kept in one piece.

A British Gas Council spokesman said: "There are no reservations on the safety of the methane (LNG) tanks at Canvey. The fears of the local residents are entirely groundless."

All of which is reminiscent of the fact that the day before the TETCO tank collapsed on Staten Island the residents were called hysterics.

Nobody ever wrote a book about the *Titanic* before she went down, or a gold-record ballad about the *Edmund Fitzgerald* before her sinking, or anything about the flawed cargo doors on the DC-10 before it crashed. A proliferation of words always follows disasters, hindsight being 20-20.

I don't claim to be a seer, but from what I have seen out there it is only a question of when, not if. And LNG is the catalyst that makes reform imperative.

I dislike being called a doomsayer. I wouldn't be writing this book if I thought we had passed the point of no return. There is still a lot we can do:

Somewhere, somebody has to sit behind a desk with a sign that says, The Buck Stops Here.

"The Office of Pipeline Safety and the Federal Power Commission are engaged in overlapping regulation of LNG storage safety. This has led to duplication of effort, fragmentation of responsibility and inefficient administration. . . .

Bluntly, two agencies—FPC and OPS—are doing the same job, and neither of them is doing it well."

Those words appeared in the Interstate and Foreign Commerce Subcommittee Report to the House of Representatives, Ninety-third Congress, 1974.

From the same report:

"The redundant regulation of LNG safety by OPS and FPC can have but one result: the public pays the price.

"It costs money for the government to staff the offices that duplicate each other's work. It costs money for industry to comply with duplicate requirements. The public pays the costs to government in higher taxes, and the cost to industry in higher prices.

"And there is no assurance that the public receives any more safety. . . .

"No real progress in this area can be expected until a single agency is vested with comprehensive jurisdiction—and can be subjected to the accountability that goes with it."

We now have eleven federal agencies—to say nothing of many more state and local agencies—involved in the approval of LNG storage sites. No wonder the buck stops nowhere.

We must have independent research. It must be free of emotionalism; it must have a sound technical base; and it must be free of economic or political bias.

We must stop expecting utilities to share the results of their research with us or even to tell us the truth.

In the *Final Report* prepared for the Office of Pipeline Safety, this statement appears:

"An important concern of the public that is sensitive to safety of LNG plants is the credibility of safety information that utilities provide to the public. In our interviews we heard frequent mention of instances where it was perceived that representatives of utilities had given inaccurate, incomplete, or misleading information on the safety of proposed or existing LNG plants. Most of the expertise on LNG is a captive of the industry, either by being in the possession of utility personnel, industry consultants, industry associations, or construction or insurance firms with their own vested interest in LNG facilities. . . . Individuals like Professors Fay and Wilson [from MIT and Harvard, respectively] have attempted to fill this role, but such independent experts have had a difficult time in gaining access to the large amount of information that industry possesses. Also, they have only limited time and funds to devote to such issues. Thus an important issue for the public is not only how safe *are* LNG facilities, but also whose pronouncements on safety can be believed."

ERDA (Energy Research and Development Administration) has recently made a research grant to Dr. Fay. It is hoped that this may result in at least one large-scale experimental spill of LNG on water. Not enough is known about the unpredictable and violent reaction known as flameless explosion (Chapter I) which occurs when the two liquids come into contact with each other. We cannot evaluate with any accuracy the rate and extent of vapor cloud (plume) travel until we know how a large quantity of LNG behaves on water, nor can we know the LNG vessel's resistance to brittle fracture when floating in its own supercold juices. There are a dozen other factors that cannot be answered satisfactorily by extrapolations from previous small experimental spills.

Certain cost-effective options may be opening up. MIT scientists are experimenting with the gel form of LNG. When small amounts of water and methyl alcohol are incorporated in liquefied natural gas, rapidly condensed and frozen, a semi-solid gel form results. If it can remain rigid under stress, and if certain engineering and economic problems can be solved, the gel form may be safer to store than the liquid.

Strong international control is essential to enforce high standards for crew training and licensing, together with frequent, thorough marine inspections. If agreement proves to be impossible it must be unilaterally imposed. Somebody, somewhere along the line, has got to get tough. Perhaps it will have to be the United States. What is the sense in building our ships to the highest standards, inspecting them, training and licensing our crews if ninety-five per cent of the maritime world uses our ports and is not subject to like requirements?

The muddy waters of flag-of-convenience shipping must somehow be clarified and controlled. But as Senate Commerce Committee Chairman Warren Magnuson said early in 1977, "I don't see how you can have control when you have American-owned ships insured by the British, run by the Greeks, with Italian officers and a Chinese crew."

Site selection for storage tanks is the most controversial point in the whole LNG wrangle. Three powerful and distinctly different camps converge on this and conflict bitterly: the "old" environmentalists, the "new" environmentalists, and industry. Each camp has its reasons and its rights.

The old environmentalists insist that the tanks be sited in industrial locations to avoid disruption of environmentally unspoiled areas. It is out of respect to them—or fear of them—that industry's impact statements have devoted literally hundreds of pages to concern for flora and fauna that the average person never heard of.

At Point Conception, California, for example, the traditional environmentalists opposed the proposed LNG terminal, not as a safety hazard to people, for only five hundred persons live within the surrounding ten square miles, but because of the threat to bird, plant, and marine life, and also its pristine beauty. A successful gas operation would mean a dependable fuel source which would attract other environment-polluting industry.

The Sierra Club is typical of the old environmentalists. Founded in 1882 by the esteemed naturalist John Muir, this group has protested and often prevented the rape of earth, air, water, and endangered species. Well organized, they have significant political and financial resources.

The new environmentalists are apt to be operating on a shoestring. These groups have sprung up suddenly in places like Rossville where they feared that the storage tanks and the ships moving in and out pose a threat, not to plankton, but to their property, their own people.

An instant change-over from old to new does sometimes take place as in Los Angeles, stronghold of the Sierra Club, when the *Sansinena* blew up just before Christmas, 1976, in their harbor. *Their* windows were knocked out; *their* Christmas trees came skittering across the floor; some of *their* kids were out in the harbor on water skis.

It is at times like this that people start writing letters to BLAST.

From Canvey Island: "Our papers over here had a story about you. You're way ahead of us. Can you send information about what your organization is doing?"

From Newport, Oregon, one of the most beautiful little harbors on the northwest coast, with winter population 5,000 and ten times that in the summer: "We are trying to investigate permissions responsible for a newly erected ten million gallon LNG tank, but three years of relevant files have disappeared from the office of the city clerk. We're raising a war chest. Can we keep Northwest Natural Gas from filling the tank? Please let us hear from you."

And the Cosgriffs always answer. Because their typewriter is broken, they write their many-paged letters by hand. They dig into their files and Xerox copies of the story about John Quinn's house five hundred feet from the tanks, together with copies of the report stating that officials of an LNG plant in Indonesia were appalled when they learned about the close proximity of the tanks to that Staten Island house. The plant in Brunei, on the Indonesian island of Sumatra, is located 2,500 feet away from the nearest building, which is not a dwelling but an administration building. Some 800 families were relocated still farther away.

To say that both old and new environmentalists have thrown gravel in the gears of industry would be the understatement of the decade. Since there is nothing on the books, no laws to keep industry from making siting decisions which are dictated only by the economics of project requirements, some companies have built tanks costing hundreds of millions in whatever location they chose.

But certain responsible members of industry have come to the conclusion that the choice of storage and dock locations should not be dictated by purely economic considerations. El Paso LNG, the world's leading importer of liquefied natural gas, with nine new 125,000-cubic-meter LNGCs and approval pending for construction of eleven more with a capacity of 165,000 cubic meters each, has proved that siting with consideration for environmental ideas, both old and new, is economically feasible. They are proud of their choice of sites at Cove Point, Maryland, and Elba Island, Georgia.

Ivan W. Schmitt, vice-president of that company, said, "We have concluded that the ideal marine terminal for LNG should be located in an unpopulated area in order to reduce the possibility—however remote—of exposing individuals and property to a flammable vapor cloud generated by an LNG water spill."

He said further that, by avoiding locations of unique natural value and meeting or exceeding all regulatory requirements aimed at protecting the environment, the net effect would be positive because LNG is a source of clean energy.

Assuming that sites as suitable as those selected by El Paso are not available, industry has other attractive cost-effective options:

Proven technology exists for offshore concrete islands serviced by pipelines buried under the sea bed. Two such offshore oil superports have been approved for the Gulf of Mexico. One called LOOP (Louisiana Offshore Oil Port) will be built eighteen miles off the Louisiana coast. The other, Seadock, will be situated twenty-six miles out in the Gulf. Safe berthing facilities for the largest tankers in the world and vast offshore storage facilities, all constructed within stringent environmental constraints, prove this to be one viable option for imported LNG.

Another solution involves using a floating harbor. Today this alternative is an off-the-shelf item that is capable of receiving, storing, and liquefying natural gas. With the first plant planned for use off the coast of Iran, Moss Rosenberg of Norway announced that the cost will be only an approximate sixty per cent of a comparable shore plant, and be capable of servicing half a dozen LNG tankers.

Furthermore, this floating harbor will require a year less construction and start-up time than comparable shore facilities. The built-in mobility of the floating unit provides the necessary flexibility to accommodate changing fuel consumption patterns.

Am I saying that these installations are accident-proof? By no means. These facilities will be just as prone to accidents as their shore-based counterparts. But basic to these solutions is safety increased to a level that people can live with.

At last the concept would become an acceptable, insurable risk.

Finally, the sabotage potential would be enormously lessened.

The behemoths would no longer have to enter our rivers and harbors.

BLAST and its counterparts could be phased out.

And the albatross of human error would hang less heavily about our necks.

Get it out of town.

BIBLIOGRAPHY

CHAPTER I

Allan, D., Atallah, S., Drake, E., Hinckley, R., and Mathias, S. *Final Report, Technology and Current Practices for Processing, Transferring, and Storing Liquefied Natural Gas.* Office of Pipeline Safety. Washington: Government Printing Office, 1974.

Condon, George E. "Has It Been 20 Years?" Cleveland *Plain Dealer,* October 20, 1964.

Curt, Robert P., and Delaney, Timothy D. *Marine Transportation of Liquefied Natural Gas.* Kings Point, New York: National Maritime Research Center, 1973.

Fay, James M., and MacKenzie, James J. "Cold Cargo," *Environment,* November 11, 1972.

Liquefied Natural Gas, Views and Practices, Policy and Safety, CG-478, February 1, 1976. Department of Transportation, U. S. Coast Guard.

Report on the Outline of Collision Between Japanese Tanker Yuyo Maru, No. 10, and Liberian Freighter Pacific Ares. Maritime Safety Agency, Japanese Government. Atlantic International Air and Surface Search and Rescue Seminar, April 1975, U. S. Coast Guard.

Van Langen, James R. *LNG Liquefaction Plants and Associated Shoreside Operations.* Kings Point, New York: National Maritime Research Center, 1973.

World Book Encyclopedia, Vol. IV, p. 102. Chicago: Field Enterprises Educational Corporation, 1963.

CHAPTER II

Brown, Thomas M., and Clark, Cheryl. "Can This Man Clean Up California?" *New York Times Magazine,* October 17, 1976.

Chernow, Ron. "Deadly Gas Threatens City," *Village Voice,* June 28, 1976.

Federal Power Commission in re: Distrigas Corporation et al.: Docket #CP-73-132 et al.; Eascogas LNG, Inc., et al.: Docket #CP-73-47 et al.

Grand Jury Report, Supreme Court of the State of New York, County of Richmond, The People of the State of New York Against John Doe #2, 1973.

Harrigan, Peter. "Scientist Warns on Tank Spills," *Staten Island Advance,* February 10, 1973.

Hill, Gladwin. "Controversy over Liquefied Natural Gas Pits Energy Needs Against Danger," New York *Times,* October 7, 1976.

Maio, Anthony. "LNG . . . An Unwelcome Neighbor," Greenpoint Gazette, April 8, 1975.

Metz, Tim. "Gas Shortages Give New Ammunition to LNG Advocates," *Wall Street Journal,* January 14, 1977.

Richards, Dan. "Expert Predicts LNG Disaster," *Staten Island Register,* May 8, 1975.

Sherrill, Robert. "Breaking Up Big Oil," *New York Times Magazine,* October 3, 1976.

Spiegel, Claire. "The LNG Story," New York *Sunday News,* January 19, 1975.

———. "Liquid Gas: Blessing or Curse?" New York *Sunday News,* December 28, 1975.

Staten Island Explosion: Safety Issues Concerning LNG Storage Facilities. Hearings Before the Special Subcommittee on Investigations of the Committee on Interstate and Foreign Commerce, House of Representatives, July 10, 11, and 12, 1973. Washington: U. S. Government Printing Office.

"Storage Tank Dispute on Staten Island Persists," New York *Times,* January 26, 1975.

Waters, Mike. "FPC Bares Loss of Gas Tank Data," Washington *Post,* July 13, 1973.

CHAPTER IV

Cockburn, Alexander, and Ridgeway, James. "Liquid Natural Gas Tankers Rouse Fear of Catastrophe," *Parade*, February 20, 1977.

"Efficient Insulation Plays a Vital Part in LNG Transport," *Marine Engineering/Log*, September 1976.

Federal Power Commission hearings, as cited for Chapter II.

"Gotaas Larsen Adds Six 125,000 cubic meter LNG Carriers to its Shipping Fleet," *Marine Engineering/Log*, September 1976.

"Japanese Fish Boats Block an A-Ship," New York *Times*, August 25, 1974.

"Japan's Nuclear Ship Returns to Port," New York *Times*, October 16, 1974.

Mostert, Noël. *Supership.* New York: Knopf, 1973.

"New Membrane Containment Systems for LNG Developed by French and U.S. Firms," *Marine Engineering/Log*, September 1976.

"Nuclear Ship Due at Mutsu," Baltimore *Sun*, October 15, 1974.

"Rice Balls Don't Seal Reactor," Seattle *Post Intelligencer*, September 9, 1974.

Seiden, Matthew J. "Nuclear Ship Faces Battle Going to Port," Baltimore *Sun*, September 18, 1974.

Tanker Advisory Center Report. New York: July 3, 1975.

CHAPTER V

Bolton, Clint. "Corps Is Paying $80,000 per Day to Dredge Pass," New Orleans *Times Picayune*, March 3, 1974.

Coast Guard Marine Casualty Report. USCG/NTSB, March 2, 1976.

Halvarsen, Fred H. "A Review of Some Recent Accidents in the Marine Transportation Mode," *Proceedings of the Marine Safety Council*, June 1975.

Ireland, George F. "For Want of a Nail," *Proceedings of the Marine Safety Council*, November 1975.

Marine Casualty Report SS CV Sea Witch-SS Esso Brussels, Collision and Fire in New York Harbor on 2 June, 1973 with Loss of Life. Report No. USCG/NTSB, December 17, 1975. Washington: National Transportation Safety Board.

Perry, James M. "Here Come the Combustible Gas Tankers," *National Observer*, February 19, 1977.

"Sea Witch-Esso Brussels." *Proceedings of Marine Safety Council*, May 1976.

"Valdez Hazards Worry Navigators," *American Maritime Officer*, January 1977.

CHAPTER VI

"A Bad Month for Liberian Tankers," *American Maritime Officer*, January 1977.

Cockburn, Alexander, and Ridgeway, James. "Fahrenheit Minus 255: The Story of a Disaster, a Deal, and a Billion Dollar Business," *Village Voice*, December 12, 1976.

"Demolition Derby at Sea," *Time*, January 24, 1977.

Kifner, John. "Liberia: A Phantom Maritime Power Whose Fleet Is Steered by Big Business," New York *Times*, February 14, 1977.

"Liberian Flag Tanker Named as Source of Florida Oil Spill," *American Maritime Officer*, December 1975.

Robards, Terry. "Burmah Oil's U.S. Aid Bid Studied for Possible Fraud," New York *Times*, August 19, 1976.

"The Lush Era of the Tanker Tycoons," Newsweek, October 19, 1970.

Ulman, Neil. "How the Arcane World of the Argo Merchant, Other Tankers Works." *Wall Street Journal*, January 18, 1977.

Villiers, Alan. *Men, Ships, and the Sea*. Washington: National Geographic Society, 1962.

———. *Posted Missing*. New York: Charles Scribner's Sons, 1974.

CHAPTER VII

Benkert, W. M., and Hill, R. C. "U. S. Coast Guard Traffic Systems," *Proceedings of the Marine Safety Council*, July 1972.

"Big Spending Sailors," *Time*, October 25, 1976.

"Collision Under the Golden Gate," *Proceedings of the Marine Safety Council*, November 1971.

Gilmore, Grant, and Black, C. L. *The Law of the Admiralty*. Brooklyn: Foundation Press, Inc., 1957.

Hearings Before the Subcommittee on Coast Guard and Naviga-

tion (Part II) of the Committee on Merchant Marine and Fisheries, House of Representatives, 93rd Congress. Washington: U. S. Government Printing Office, 1975.

Joyce, Benjamin E. "Men at Sea—Then and Now," *Proceedings of the Marine Safety Council,* January 1976.

Keylin, Arleen, and Brown, Gene, editors. *Disasters from the Pages of the New York Times.* New York: Arno Press, 1976.

Lord, Walter. "An Epic Sea Rescue," *Life,* August 6, 1956.

Moran, Hank. "Safety: We Are the Enemy," *Proceedings of the Marine Safety Council,* April 1974.

Moscow, Alvin. *Collision Course.* New York: Putnam, 1959.

Mostert, Noël, as cited for Chapter IV.

Nature of Ship Collisions Within Ports. U. S. Department of Commerce National Technical Information Service, PB-255-304, April 1956.

Schwimmer, Martin J. *Environmental Analysis: A Framework for Improving Efficiency of Maritime Personnel.* National Maritime Research Center, Kings Point, New York.

U. S. Seamen and the Seafaring Environment Symposium Report. Washington: U. S. Department of Commerce, Maritime Administration.

White, Peter. "Barehanded Battle to Cleanse the Bay," *National Geographic,* June 1971.

CHAPTER VIII

"Chronicle of Disaster," *Proceedings of the Marine Safety Council,* February 1972.

Fennell, James E. "Ghost Ships Are Menace to World's Ocean Lanes," *American Maritime Engineering,* September 1974.

"Final Fitzgerald Report Now Slated for Spring," *American Maritime Officer,* February 1977.

Gilles, Audy. "200,000 cubic meter LNG Tankers," *Ocean Industry,* November 1972.

Sherrill, Robert. "The Cheap Door That Cost 346 Lives," *New York Times Book Review,* October 10, 1976.

Structural Failure and Sinking of the Texaco Oklahoma off Cape Hatteras on 27 March, 1971, with the Loss of 31 Lives. U. S.

Coast Guard Marine Board of Investigation Report and Commandant's Action. Washington: Department of Transportation.

CHAPTER IX

"Captain Prays after Liberian-Flag Voyage," *American Maritime Officer*, March 1977.

"Chronicle of Disaster," *Proceedings of the Marine Safety Council*, February 1972.

Drake, Elisabeth, and Reid, Robert C. "The Importation of Liquefied Natural Gas," *Scientific American*, April 1977.

Ingram, Timothy H. "Peril of the Month: Gas Supertankers," *Washington Monthly*, February 1973.

Joyce, Benjamin E., as cited for Chapter VII.

Moscow, Alvin, as cited for Chapter VII.

Nature of Ship Collisions Within Ports. Prepared by Todd Shipyards Corporation for National Maritime Research Center, Galveston, April 1976, U. S. Department of Commerce, National Technical Information Center PB 255–304.

Van Horn, Andrew J., and Wilson, Richard. *Liquefied Natural Gas: Safety Issues, Public Concerns and Decision Making*. Cambridge, Mass.: Energy and Environmental Policy Center, Harvard University, 1976.

CHAPTER X

"Behind the Siege of Terror in Washington," *U. S. News and World Report*, March 21, 1977.

Flood, Michael. "Nuclear Sabotage," *Bulletin of the Atomic Scientists*, September 1976.

"Hijacked Plane Lands in Havana . . . ," New York *Times*, November 12, 1972.

"Latest Worry: Terrorists Using High Technology," *U. S. News and World Report*, March 14, 1977.

"Man with Razor Seized in Jet Hijack Attempt," *Daily Progress*, May 9, 1977.

"Manson and the Presidency," editorial, *National Review*, September 26, 1975.

Phillips, John Aristotle. "How I Designed the A-Bomb," *Science Digest*, January 1977.

"Seizing Hostages: Scourge of the '70's," *Newsweek*, March 21, 1977.

"Terror in Washington," *U. S. News and World Report*, March 21, 1977.

Turner, Wallace. "The Gun Is Pointed . . . ," New York *Times*, September 12, 1975.

Van Horn, Andrew J., and Wilson, Richard, as cited for Chapter IX.

CHAPTER XI

Allan, D., et al., *Final Report* as cited for Chapter I.

Ferguson, Peter. "Gas Men Abandon Canvey Island Earth Pit Fight," *New Civil Engineer*, November 20, 1975.

Grey, Michael. "Kicking the Political Football at Gastech," *Fairplay International Shipping Weekly*, October 14, 1976.

Jacobson, Philip. "Is This the Scenario for a New Disaster?" London *Sunday Times*, September 6, 1974.

Kress, Michael E., Vachtsevanos, George J., and Leonard, Brian P. *LNG Transportation and Storage: Relative Risks, Hazards and Alternatives*. Division of Pure and Applied Sciences, Richmond College, City University of New York, September 1974.

Legislative Issues Relating to the Safety of Liquefied Natural Gas. Report by the Special Subcommittee on Investigations of the Committee on Interstate and Foreign Commerce, House of Representatives, 93rd Congress, 1974. Washington: U. S. Government Printing Office.

Ratcliff, Spencer. "The Rape of an Essex Island," *Evening Echo*, April 17, 1973.

Schmitt, Ivan W. "The Trans-Alaska Gas Project Site Selection and System Optimization," *Marine Technology*, January 1976.

Turner, Norma. "Will Canvey be the Next Flixborough?" *New Scientist*, November 27, 1975.

Van Horn, Andrew J., and Wilson, Richard, as cited for Chapter IX.

Wilson, Richard. Asilomar Conference on Risk/Benefit Analysis . . . Session on LNG, September 26, 1975.